SRA Connecting Math Concepts

Level D Textbook

COMPREHENSIVE EDITION

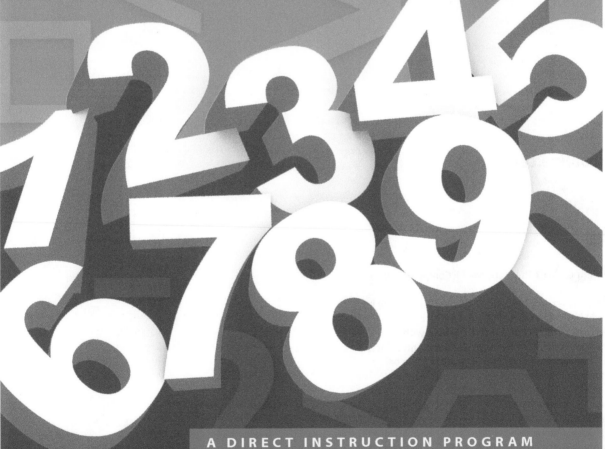

A DIRECT INSTRUCTION PROGRAM

McGraw Hill Education

Bothell, WA • Chicago, IL • Columbus, OH • New York, NY

MHEonline.com

Send all inquiries to:
McGraw-Hill Education
8787 Orion Place
Columbus, OH 43240

ISBN: 978-0-02-103632-5
MHID: 0-02-103632-2

Printed in the United States of America.

3 4 5 6 7 8 9 QVS 15 14 13

Lesson

		15		Name	

P1

a. 3 4 9
   + 1 2 2

b. 2 2 9
   + 1 3 3

c. 2 7 1
   + 3 7 8

P2

a. 5 10 ▪ ▪ ▪ ▪ ▪ ▪ ▪ 50

b. 2 4 ▪ ▪ ▪ ▪ ▪ ▪ ▪ 20

P3

a. 70 + 10 = ▪

b. 40 + 10 = ▪

c. 75 + 10 = ▪

d. 45 + 10 = ▪

Lesson 16

		16	Name
P1			
a.	488 +258	b. 374 −161	c. 388 +143
P2			
a.	5 10 ■ ■ ■ ■ ■ ■ ■ 50		
b.	2 4 ■ ■ ■ ■ ■ ■ ■ 20		
P3			
a.	37 + 10 = ■	b. 43 + 10 = ■	c. 49 + 10 = ■
P4			
a.	5 × 7 = ■	b. 9 × 4 = ■	c. 10 × 3 = ■

Connecting Math Concepts

Lesson

Part 1

a. Jerry has 27 more pencils than **T**ony has.
   Tony has 12 pencils.
   How many pencils does Jerry have?

b. **D**avid weighed 21 pounds less than **B**ob weighed.
   Bob weighed 45 pounds.
   How many pounds did David weigh?

c. **F**rank bought 24 more apples than **A**nn bought.
   Ann bought 11 apples.
   How many apples did Frank buy?

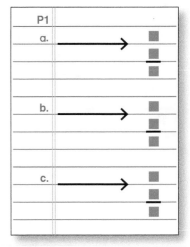

Independent Work

Part 2 For each shape, write a count-by problem and figure out the number of squares.

a.

b.

c.

P2			
a. ■ × ■ = ■	b. ■ × ■ = ■	c. ■ × ■ = ■

Part 3 Copy each problem and write the answer.

P3			
a. 20 + 30 = ■	b. 40 + 40 = ■	c. 30 + 30 = ■
d. 40 + 20 = ■	e. 20 + 50 = ■	f. 60 + 30 = ■

Lesson 17

Part 4 Write the place-value fact for each number.

P4	
a. ▪ + ▪ + ▪ = 341	b. ▪ + ▪ + ▪ = 217

Part 5 Copy each problem and work it.

P5				
a. $\begin{array}{r} 276 \\ +386 \\ \hline \end{array}$	b. $\begin{array}{r} 385 \\ -21 \\ \hline \end{array}$	c. $\begin{array}{r} 269 \\ +160 \\ \hline \end{array}$	d. $\begin{array}{r} 35 \\ +95 \\ \hline \end{array}$

Part 6 Write the numbers for counting by 9s to 90.

P6	
a. 9 18 ▪ ▪ ▪ ▪ ▪ ▪ ▪ 90	

Lesson

Part 1

a. **S**ally made 6 fewer cakes than **M**olly made.
Sally made 20 cakes.
How many cakes did Molly make?

b. **J**erry had 12 fewer marbles than **M**elissa had.
Melissa had 26 marbles.
How many marbles did Jerry have?

c. **S**am found 18 more seashells than **G**ary found.
Gary found 11 seashells.
How many seashells did Sam find?

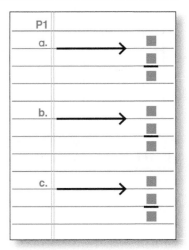

Part 2

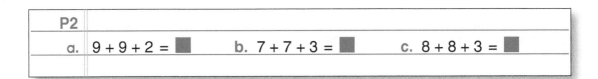

P2			
a.	$9 + 9 + 2 = $ ■	b. $7 + 7 + 3 = $ ■	c. $8 + 8 + 3 = $ ■

Independent Work

Part 3 Copy each number family and write the big number.
Then write 2 addition and 2 subtraction facts for each family.

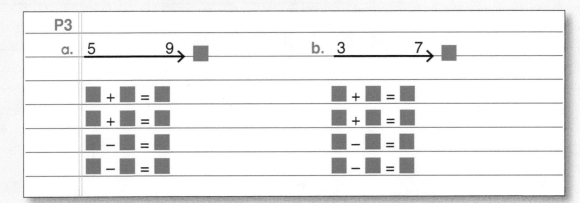

Lesson 18

Part 4 For each shape, write a count-by problem and figure out the number of squares.

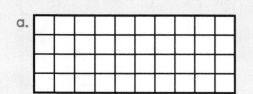

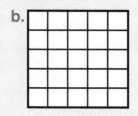

P4		
a.	■ × ■ = ■	b. ■ × ■ = ■

Part 5 Copy each problem and write the answer.

P5			
a.	30 + 40 = ■	b. 20 + 30 = ■	c. 50 + 20 = ■
d.	50 + 10 = ■	e. 40 + 40 = ■	f. 30 + 30 = ■

Part 6 Copy and work each problem.

P6				
a.	392 + 592	b. 498 − 107	c. 389 + 65	d. 417 + 307

Part 7 Write the numbers for counting by 9s to 90.

P7	
a.	9 18 ■ ■ ■ ■ ■ ■ ■ 90

Lesson 19

Part 1

a. The **c**at was 8 pounds lighter than the **d**og.
The cat weighed 11 pounds.
How many pounds did the dog weigh?

b. **G**ary was 12 years older than **H**elen.
Gary was 56 years old.
How many years old was Helen?

c. **K**en had 11 fewer crayons than **D**avid had.
Ken had 33 crayons.
How many crayons did David have?

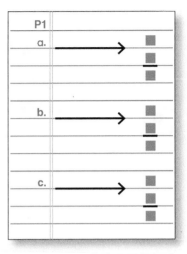

Part 2

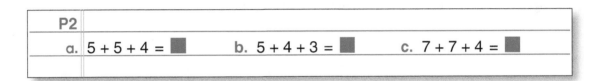

a. 5 + 5 + 4 = ■   b. 5 + 4 + 3 = ■   c. 7 + 7 + 4 = ■

Independent Work

Part 3 Copy the numbers in each item and write the sign >, <, or =.

P3			
a.	245 ■ 427	b. 328 ■ 394	c. 294 ■ 294

Part 4 For each shape, write a count-by problem and figure out the number of squares.

a. 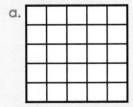   b. 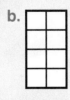   c.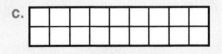

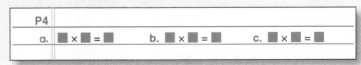

P4			
a.	■ × ■ = ■	b. ■ × ■ = ■	c. ■ × ■ = ■

Lesson 19

Part 5 | Write the numbers for counting by 4s to 40.

P5	
a.	4 ■ ■ ■ ■ ■ ■ ■ ■ 40

Part 6 | Copy each problem and work it.

P6	

a.   4 5 7
   + 2 6 7
	b. 9 3 7
	   − 8 0 6 c. 2 4 5
	   + 6 5 d. 8 9
	   + 3 2

Part 7 | Copy each number family and write the big number.
Then write 2 addition and 2 subtraction facts for each family.

P7	
a. 3 ⟶ 9 , ■ b. 4 ⟶ 7 , ■	
■ + ■ = ■ ■ + ■ = ■	
■ + ■ = ■ ■ + ■ = ■	
■ − ■ = ■ ■ − ■ = ■	
■ − ■ = ■ ■ − ■ = ■	

Lesson 20

Part 1

a. Ann is 31 years old.
Ann is 7 years younger than Kathy.
How many years old is Kathy?

b. Bob is 15 pounds heavier than Ann.
Ann weighs 119 pounds.
How many pounds does Bob weigh?

c. Harry has 48 stickers.
Kathy has 23 fewer stickers than Harry.
How many stickers does Kathy have?

d. Bob is 14 years older than Kelly.
Kelly is 26 years old.
How many years old is Bob?

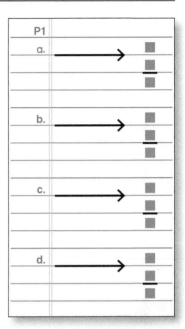

Independent Work

Part 2

a. 55 60 ■ 70
b. 298 297 ■ 295
c. 42 44 ■ 48
d. 85 90 ■ 100

P2			
a.	b.	c.	d.

Part 3 Copy the numbers in each item and write the sign >, <, or =.

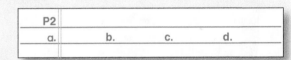

P3			
a. 245 ■ 427	b. 328 ■ 394	c. 294 ■ 294	

Part 4 Copy each problem and work it.

P4			
a. 485 +295	b. 495 −285	c. 362 +164	d. 284 −170

Lesson 20

Part 5 For each shape, write a count-by problem and figure out the number of squares.

a.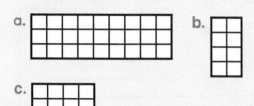

b.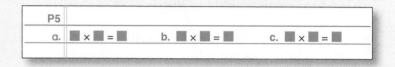

c.

P5			
a. ■ × ■ = ■	b. ■ × ■ = ■	c. ■ × ■ = ■	

Connecting Math Concepts

Lesson 21

Part 1

a. The red truck is 41 feet long.
The red truck is 12 feet shorter than the blue truck.
How many feet long is the blue truck?

b. The big tent is 19 years older than the small tent.
The small tent is 2 years old.
How many years old is the big tent?

c. The brick house is 35 feet tall.
The brick house is 10 feet taller than the wood house.
How many feet tall is the wood house?

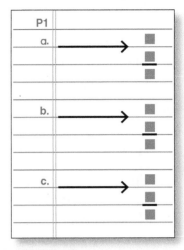

Independent Work

Part 2 Copy and work each problem.

	P2					
a.		3 2 4	b.	2 2 4	c.	2 9 7
		1 5 2		+ 1 3 4		+ 5 7
		+ 2 2 6				

Part 3 Copy and work each problem.

P3

a. $9 \times 5 = $ ▪   b. $5 \times 4 = $ ▪   c. $4 \times 6 = $ ▪

Part 4 Write the missing number for each item.

a. 52 54 ▪ 58   b. 35 40 ▪ 50

c. 87 86 ▪ 84   d. 341 342 ▪ 344

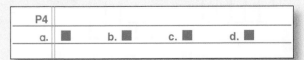

Lesson 22

a. The redwood tree is 120 feet tall.
   The redwood tree is 20 feet taller than the maple tree.
   How many feet tall is the maple tree?

b. Tina walked 53 miles.
   Alicia walked 22 fewer miles than Tina walked.
   How many miles did Alicia walk?

c. The cow weighs 124 pounds less than the horse.
   The cow weighs 819 pounds.
   How many pounds does the horse weigh?

d. Jerry read 85 books last year.
   Jerry read 17 fewer books than Mario read.
   How many books did Mario read?

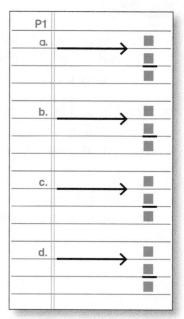

Independent Work

Part 2 Copy each number family and write the big number.
Then write 2 addition and 2 subtraction facts for each family.

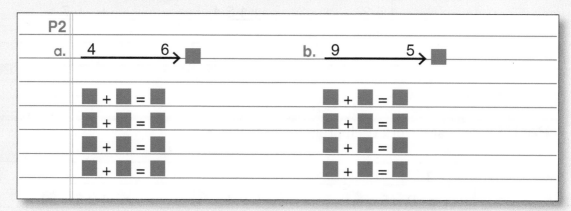

Part 3 For each shape, write a count-by problem and figure out the number of squares.

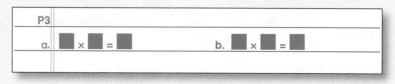

Lesson

Part 1

a. Don bought 3 pencils.

b. Jan gave away 12 pencils.

c. He lost 2 books.

d. Tim found 4 books.

P1	
a.	⟶
b.	⟶
c.	⟶
d.	⟶

Part 2

a.

b.

c.

d.

P2			
a.	b.	c.	d.

Independent Work

Part 3

a. Jerry had 14 fewer books than Ted had.
Jerry had 46 books.
How many books did Ted have?

b. The blue house is 149 years old.
The blue house is 26 years older than the green house.
How many years old is the green house?

c. The horse is 35 pounds heavier than the cow.
The horse weighs 865 pounds.
How many pounds does the cow weigh?

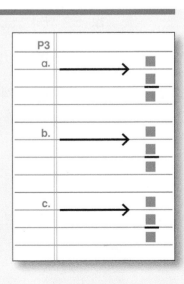

Lesson 23

Part 4 Write the answer to each question.

a. How many feet equals 1 yard?

b. How many hours equals 1 day?

c. How many minutes equals 1 hour?

d. How many seconds equals 1 minute?

e. How many cents equals 1 dollar?

P4				
a.	b.	c.	d.	e.

Part 5 Copy the numbers in each item and write the sign >, <, or =.

P5	
a. 284 ■ 192	b. 306 ■ 360

Part 6 Copy and work each problem.

P6		
a. 349 + 52	b. 278 −170	c. 85 35 +25

Part 7 Write the missing number for each item.

a. 80 85 90 ■ 100

b. 52 54 56 ■ 60

c. 98 97 96 ■ 94

d. 54 64 74 ■ 94

P7				
a. ■	b. ■	c. ■	d. ■	

Lesson 24

Part 1

a. A parking lot started with 13 cars in it.
Then 32 more cars drove into the lot.
How many cars did the lot end with?

b. The farmer started with 48 melons.
Then she sold 23 melons.
How many melons did the farmer end with?

c. The bus started out with 18 people on it.
Then 21 more people got on the bus.
How many people ended up on the bus?

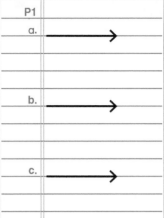

P1	
a.	→
b.	→
c.	→

Part 2

a.

b.

c.

P2		
a.	b.	c.

Lesson  24

Part 3

a. Tom read 154 pages.
Tom read 42 fewer pages than Albert read.
How many pages did Albert read?

b. The couch is 35 pounds lighter than the bed.
The couch weighs 185 pounds.
How many pounds does the bed weigh?

c. Jerry has 125 stickers.
Molly has 78 more stickers than Jerry has.
How many stickers does Molly have?

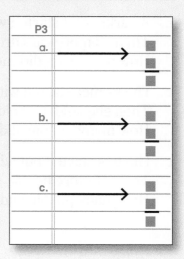

Part 4 | Write the answer to each question.

a. How many feet equals 1 yard?

b. How many inches equals 1 foot?

c. How many days equals 1 week?

d. How many seconds equals 1 minute?

e. How many hours equals 1 day?

f. How many cents equals 1 dollar?

P4					
a.	b.	c.	d.	e.	f.

Part 5 | Copy each problem and write the answer.

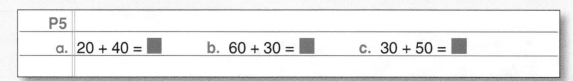

P5			
a. 20 + 40 = ■	b. 60 + 30 = ■	c. 30 + 50 = ■

Lesson 25

Part 1

a. The farmer started with 33 dogs.
He gave away 13 dogs.
How many dogs did he end with?

b. Mike started out with 13 video games.
Then he bought 6 more video games.
How many games did he end up with?

c. The man started with 134 strawberries.
He ate 21 strawberries.
How many strawberries did he end with?

P1	
a.	→
b.	→
c.	→

Part 2

a. How many pounds did John weigh? 139

b. How many years did the tree live? 77

c. How many coins did Carla have? 12

d. How many days did it rain? 17

e. How many inches long was the snake? 39

f. How many towns did they visit? 7

P2	
a.	
b.	
c.	
d.	
e.	
f.	

Independent Work

Part 3

a. The big horse is 185 pounds heavier than the little horse.
The little horse weighs 354 pounds.
How many pounds does the big horse weigh?

b. The tree is 90 feet tall.
The tree is 64 feet taller than the house.
How tall is the house?

c. Randy is 24 years younger than Alex.
Alex is 60 years old.
How old is Randy?

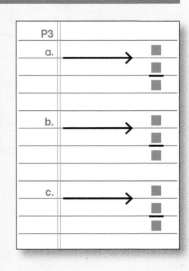

Lesson 25

Part 4 For each item, write the column problem to find the missing number.

a. $24 + \blacksquare = 90$ b. $114 + \blacksquare = 195$

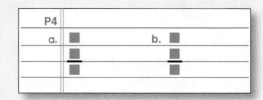

Part 5 Write the missing number for each item.

a. 29   39   ■   59   b. 98   97   ■   95

c. 54   64   ■   84   d. 412   413   ■   415

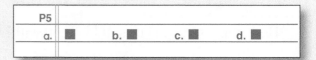

Part 6 Write the missing number for each fact.

a. ■ cents equals 1 quarter   d. ■ hours equals 1 day

b. ■ days equals 1 week   e. ■ inches equals 1 foot

c. ■ seconds equals 1 minute   f. ■ feet equals 1 yard

P6		
a.	b.	c.
d.	e.	f.

Part 7 Copy the number family and write the big number.
Then write 2 addition and 2 subtraction facts.

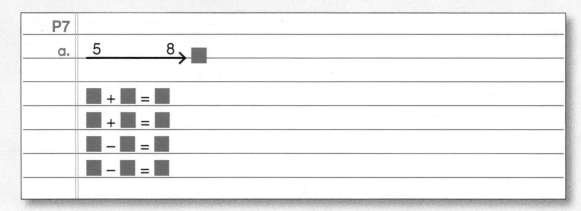

Lesson 25

Part 8

a.

b.

c.

P8		
a.	b.	c.

Part 9 Copy and work each problem.

P9			
a.	3 2 8	b.	9 4
	1 5 1		1 3 5
	+ 2 5 2		+ 2 3

Lesson 26

Part 1

a. Tom had 48 cherries.
   He ate 15 cherries.
   How many cherries did he end up with?

b. There were 38 people on the bus.
   Then 12 people got off the bus.
   How many people ended up on the bus?

c. There were 25 people on a plane.
   Then 42 people got on the plane.
   How many people ended up on the plane?

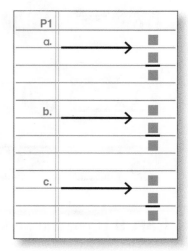

Part 2

P2			
a.	$ 2. 1 3	b. $ 4. 3 6	c. $ 2. 4 5
	+ 1. 0 5	+ 5. 9 0	− 1. 3 2

Part 3

a. How many gallons did the tank hold? 12

b. How many feet taller is the red house? 6

c. How many legs did the bug have? 8

d. How many girls were in the house? 12

e. How many hours did they drive? 5

f. How many quarters did she have? 28

P3	
a.	
b.	
c.	
d.	
e.	
f.	

Lesson

Part 4

a. Tom had 14 fewer pencils than Bob had.
Bob had 35 pencils.
How many pencils did Tom have?

b. Karen is 50 years old.
Karen is 12 years younger than Jane.
How many years old is Jane?

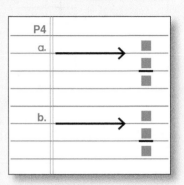

Part 5 Copy each problem and work it.

	P5			
a.	403	b. 590	c. 715	d. 607
	−172	−465	− 95	− 56

Part 6 Write the missing number for each fact.

a. ■ feet equals 1 yard

b. ■ seconds equals 1 minute

c. ■ hours equals 1 day

d. ■ months equals 1 year

e. ■ inches equals 1 foot

P6				
a.	b.	c.	d.	e.

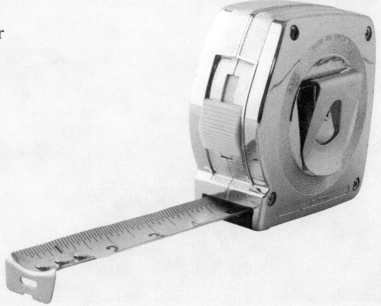

Lesson 26

Part 7 For each shape, write a count-by problem and figure out the number of squares.

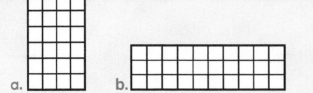

P7		
a.		b.

Part 8 Write the numbers for counting by 4s to 40.

P8	
a.	4 ▪ ▪ ▪ ▪ ▪ ▪ ▪ 40

Lesson 27

Part 1

a. Bill started out with some marbles.
Then he gave away 10 marbles.
He ended up with 47 marbles.
How many marbles did he start out with?

b. There were some passengers on a bus.
Then 11 passengers got on the bus.
The bus ended up with 14 passengers.
How many passengers started out on the bus?

c. John had some peanuts.
He ate 14 peanuts.
He ended up with 25 peanuts.
How many peanuts did he start out with?

P1	
a.	→ ■ ▬ ■
b.	→ ■ ▬ ■
c.	→ ■ ▬ ■

Part 2

< $2.20   < $12.50   < $2.99   < $7.05

a. A person buys the cup and the scarf.

b. A person buys the cup and the notebook.

c. A person buys the scarf and the pencils.

P2			
a. ■ ▬ ■		b. ■ ▬ ■	c. ■ ▬ ■

Independent Work

Part 3

a. There are 85 cars in the parking lot.
There are 14 fewer trucks than cars in the parking lot.
How many trucks are in the parking lot?

b. Adam weighs 45 pounds less than Jim weighs.
Adam weighs 139 pounds.
How many pounds does Jim weigh?

P3	
a.	→ ■ ▬ ■
b.	→ ■ ▬ ■

Lesson 27

Part 4 Write the place-value fact for each number.

P4		
a. ■ + ■ + ■ = 205		b. ■ + ■ + ■ = 420
c. ■ + ■ + ■ = 417		d. ■ + ■ + ■ = 364

Part 5 Copy each problem and work it.

P5				
a. 487 −396	b. 511 +199	c. 324 +582	d. 950 −247

Part 6 Write the number of cents for each row.

P6		
a. ■ cents	b. ■ cents	c. ■ cents

Lesson 28

Part 1

a. A bus started out with some people.
Then 12 more people got on the bus.
35 people ended up on the bus.
How many people did the bus start out with?

b. A truck started out with some boxes on it.
Then the driver took 67 boxes off the truck.
The truck ended up with 21 boxes on it.
How many boxes did the truck start out with?

c. Jane started with some boxes.
She made 20 more boxes.
She ended with 32 boxes.
How many boxes did she start with?

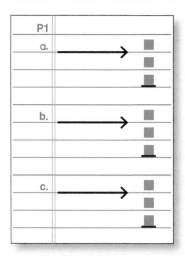

Part 2

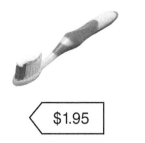   

$1.95 $5.50 $ .45 $2.19

a. You want to buy the toothbrush and the hairbrush.
How much money do you need?

b. You want to buy the hairbrush and the soap.
How much money do you need?

c. You want to buy the toothbrush and the toothpaste.
How much money do you need?

Lesson 28

Part 3 For each item, write the column problem to find the missing number.

a. 38 + ■ = 69   **b.** 39 + 27 = ■

c. 51 + ■ = 73   **d.** 59 − ■ = 13

P3				
a. ■	b. ■	c. ■	d. ■	
■	■	■	■	
■	■	■	■	

Part 4 Write the missing number for each item.

a. 160 170 ■ 190   **b.** 99 89 ■ 69

c. 47 57 ■ 77   **d.** 50 48 ■ 44

P4			
a. ■	b. ■	c. ■	d. ■

Part 5 Write 2 addition and 2 subtraction facts for this family.

a. $\xrightarrow{\hspace{0.3cm}3\hspace{1.5cm}7\hspace{0.3cm}}$

P5	
a. ■ + ■ = ■	
■ + ■ = ■	
■ − ■ = ■	
■ − ■ = ■	

Part 6 Write the dollars and cents amount for each item.

P6				
a.	b.	c.	d.	

a.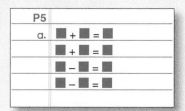

b.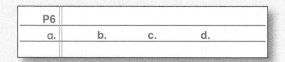

c.

d.

Lesson 28

Part 7 Copy the numbers in each item and write the sign >, <, or =.

P7			
a. 280 ■ 302	b. 421 ■ 412	c. 200 ■ 300	

Part 8 Work each problem.

a. 325 + 15 + 92   b. 146 + 205 + 51

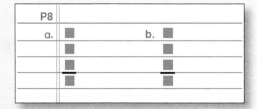

Lesson 29

Part 1

a. A truck had some chickens on it.
Then the driver put 33 more chickens on the truck.
The truck ended up with 55 chickens on it.
How many chickens did the truck start out with?

b. Millie started the day with 40 dollars.
She earned 25 dollars.
How many dollars did Millie end with?

c. At the start of the day, Rob had 68 marbles.
Then he gave away 26 marbles.
How many marbles did he end up with?

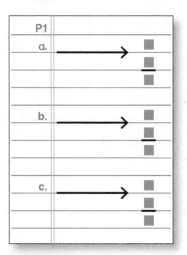

Part 2

$31.60 $44.25 $12.95 $10.00

a. You want to buy the boots and the gloves.
How much money do you need?

b. You want to buy the coat and the shirt.
How much money do you need?

Connecting Math Concepts

Lesson 29

Independent Work

Part 3 Write the answer to each question.

a. What is 40 + 30?

b. What is 64 + 10?

c. What is 20 + 50?

d. What is 70 – 20?

e. What is 64 – 10?

f. What is 93 – 10?

g. What is 80 – 30?

h. What is 40 + 40?

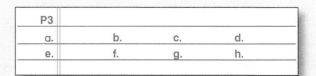

P3				
a.		b.	c.	d.
e.		f.	g.	h.

Part 4 Write the place-value fact for each number.

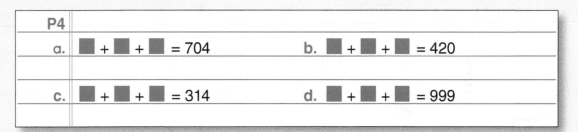

P4	
a. ▪ + ▪ + ▪ = 704	b. ▪ + ▪ + ▪ = 420
c. ▪ + ▪ + ▪ = 314	d. ▪ + ▪ + ▪ = 999

Part 5 For each item, write the column problem to find the missing number.

a. 135 + ▪ = 260

b. 36 + ▪ = 58

c. 94 – 38 = ▪

d. 72 – ▪ = 36

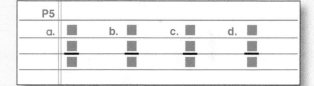

P5							
a. ▪		b. ▪		c. ▪		d. ▪
▪		▪		▪		▪

Part 6 For each shape, write a count-by problem and figure out the number of squares.

a.

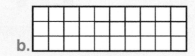

b.

P6	
a.	b.

Connecting Math Concepts

Lesson 29 **29**

Lesson 30

Part 1

a. Alicia had 13 seashells at the start of the day.
   She found 16 seashells during the day.
   How many seashells did she end up with?

b. A train had some people on it.
   Then 19 people got on the train.
   The train ended up with 89 people on it.
   How many people did the train start with?

c. There were 97 pies in the bakery at the start of the day.
   The baker sold 34 of the pies.
   How many pies were in the bakery at the end of the day?

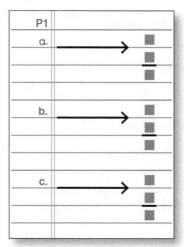

Independent Work

Part 2 Copy each problem and work it.

P2			
a. $\begin{array}{r} 540 \\ -409 \\ \hline \end{array}$	b. $\begin{array}{r} 490 \\ -184 \\ \hline \end{array}$	c. $\begin{array}{r} 690 \\ -41 \\ \hline \end{array}$	d. $\begin{array}{r} 809 \\ -53 \\ \hline \end{array}$

Part 3 Write 2 addition and 2 subtraction facts for each number family.

a. $\dfrac{7 \qquad 9}{\qquad\qquad}\!\!\longrightarrow$ b. $\dfrac{7 \qquad 5}{\qquad\qquad}\!\!\longrightarrow$

P3		
a. ■ + ■ = ■		b. ■ + ■ = ■
■ + ■ = ■		■ + ■ = ■
■ − ■ = ■		■ − ■ = ■
■ − ■ = ■		■ − ■ = ■

Part 4 Write the missing number in each row.

a. 65 60 ■ 50

b. 18 16 ■ 12

c. 37 47 ■ 67

P4		
a.	b.	c.

Lesson 30

Part 5

$1.95 $5.50 $ .45 $2.19

a. You want to buy the soap and the hairbrush.
   How much money do you need?

b. You want to buy the toothpaste and the toothbrush.
   How much money do you need?

P5		
a.	■	b. ■
	■	■

Part 6   Write the dollars and cents amount for each item.

a.

b.

P6		
a.	b.	c.

c.

Connecting Math Concepts

Lesson 31

a. A truck had ducks on it.
   Then the driver put 24 more ducks on the truck.
   The truck ended up with 55 ducks on it.
   How many ducks did the truck start out with?

b. Millie had 60 dollars.
   She found 25 dollars.
   How many dollars did Millie end with?

c. At the start of the day, Jerry had 68 marbles.
   He lost 14 marbles.
   How many marbles did he end up with?

d. The train started out with some passengers.
   125 passengers got on the train.
   The train ended up with 260 passengers.
   How many passengers did the train start with?

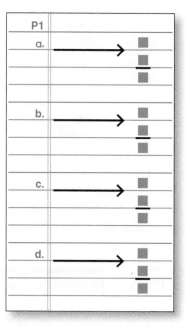

Independent Work

Part 2 | For each item, write the column problem to find the missing number.

a. 190 + ■ = 850   b. 205 + 305 = ■

c. 42 − ■ = 19   d. 180 + ■ = 300

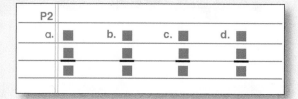

Part 3 | Write the place-value fact for each number.

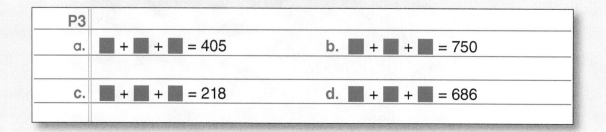

a. ■ + ■ + ■ = 405   b. ■ + ■ + ■ = 750

c. ■ + ■ + ■ = 218   d. ■ + ■ + ■ = 686

Lesson 31

Part 4 For each shape, write a count-by problem and figure out the number of squares.

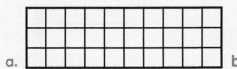

P4	
a.	b.

Part 5 Write **H** for hexagon, **P** for pentagon, **R** for rectangle, **S** for square, and **T** for triangle.

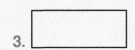

1.   2.   3.   4.

5.   6.   7.   8.

P5				
1.	2.	3.	4.
5.	6.	7.	8.

Part 6 Copy each problem and work it.

P6	
a.	$ 1 4 . 3 6
	+ 2 3 . 9 0

b. $ 7. 4 0
    + 1. 9 0

Part 7 Write the missing number for each item.

a. 185   184   183   ■   181

b. 140   130   120   ■   100

c. 862   863   864   ■   866

d. 155   160   165   ■   175

e. 36   46   56   ■   76

P7				
a.	b.	c.	d.	e.

Lesson 32

Part 1

a. Fran had 16 more books than Ted had.
   Ted had 26 books.
   How many books did Fran have?

b. The green house was 27 years older than the blue house.
   The green house was 50 years old.
   How old was the blue house?

c. Rita weighed 18 more pounds than Alice.
   Rita weighed 62 pounds.
   How many pounds did Alice weigh?

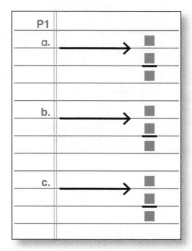

Independent Work

Part 2

For each problem, make a number family with the letters for Start and End.
Then work the problem and write the unit name.

a. Tonya had some pens.
   She bought 25 pens.
   She ended up with 61 pens.
   How many pens did she start with?

b. Jamal had some cherries.
   He ate 35 of the cherries.
   He ended up with 64 cherries.
   How many cherries did he start with?

c. Debbie started out with 85 dollars.
   She spent 19 dollars.
   How many dollars did she end up with?

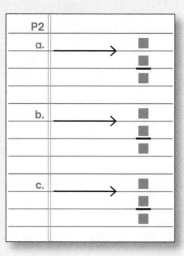

Part 3

Copy each problem and work it.

P3				
a.	487 + 97	b. 128 − 62	c. 950 −117	d. 707 − 54

Lesson 32

Part 4

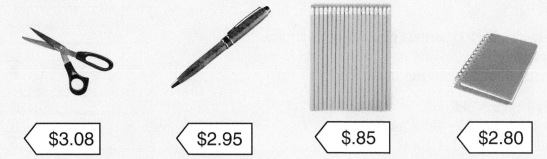

$3.08 $2.95 $.85 $2.80

a. A person buys the scissors and the pen.

b. A person buys the pencils and the notebook.

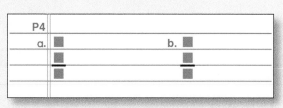

P4		
a. ■		b. ■
■		■
■		■

Part 5 Write **H** for hexagon, **P** for pentagon, **R** for rectangle, **S** for square, and **T** for triangle.

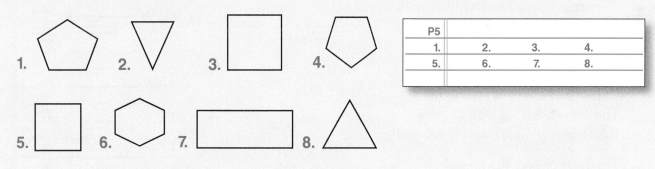

1.   2.   3.   4.

P5				
1.	2.	3.	4.
5.	6.	7.	8.

5.   6.   7.   8.

Part 6 Write the missing number for each fact.

a. ■ cents equals 1 dime

b. ■ minutes equals 1 hour

c. ■ months equals 1 year

d. ■ seconds equals 1 minute

e. ■ feet equals 1 yard

f. ■ hours equals 1 day

g. ■ cents equals 1 quarter

h. ■ days equals 1 week

i. ■ inches equals 1 foot

j. ■ cents equals 1 dollar

P6				
a.	b.	c.	d.	e.
f.	g.	h.	i.	j.

Lesson 33

Part 1

a. The cow weighs 106 pounds more than the horse.
   The cow weighs 622 pounds.
   How many pounds does the horse weigh?

b. Jerry is 49 years old.
   Tom is 12 years older than Jerry.
   How many years old is Tom?

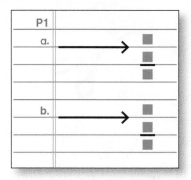

Independent Work

Part 2

For each problem, make a number family with the letters for **Start** and **End**.
Then work the problem and write the unit name.

a. The train started out with 146 passengers.
   65 passengers got on the train.
   How many passengers ended up on the train?

b. A bus started out with some people.
   25 more people got on the bus.
   The bus ended up with 90 people.
   How many people started out on the bus?

c. The farmer had 465 cows.
   The farmer sold 82 cows.
   How many cows did the farmer end up with?

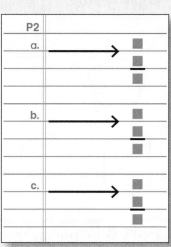

Part 3

For each item, write the column problem to find the missing number.

a. $114 + 296 = \blacksquare$

b. $455 + \blacksquare = 605$

c. $640 - 80 = \blacksquare$

d. $85 + 32 + 45 = \blacksquare$

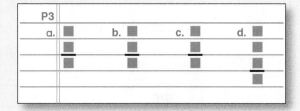

Lesson 33

Part 4 Copy each problem and work it.

P4			
a.	$ 4.26 + 1.93	b. $ 5.82 + 2.39	c. $ 7.19 + 2.69

Part 5 Write **H** for hexagon, **P** for pentagon, **R** for rectangle, **S** for square, and **T** for triangle.

1.   2.  3.   4.

P5				
1.		2.	3.	4.
5.		6.	7.	8.

5.   6. 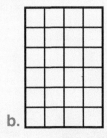 7.  8.

Part 6 For each shape, write a count-by problem and figure out the number of squares.

a. 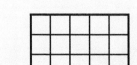 b.

P6		
a.		b.

Lesson 34

Part 1

a. Linda has 7 more notebooks than James.
James has 5 notebooks.
How many notebooks does Linda have?

b. Tom's book has 73 fewer pages than Mindy's book.
Mindy's book has 397 pages.
How many pages does Tom's book have?

c. The school has 51 fewer tables than chairs.
The school has 82 chairs.
How many tables does the school have?

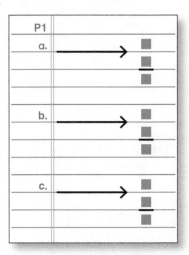

Part 2
Write **Cu** for Cube, **Py** for Pyramid, **Sp** for Sphere, **H** for Hexagon, **P** for Pentagon, **R** for Rectangle, **S** for Square, **T** for Triangle.

1.   2.   3.   4.

P2				
1.	2.	3.	4.
5.	6.	7.	8.

5.   6.   7.   8.

Lesson  34

Part 3 For each problem, make a number family with the letters for **Start** and **End**. Then work the problem and write the unit name.

a. The library had 846 books.
A man gave the library 60 more books.
How many books did the library end with?

b. The bookstore started the day with lots of books.
The bookstore sold 194 books during the day.
The bookstore ended the day with 245 books.
How many books did the bookstore start with?

c. Jerry started with some stickers.
His teacher gave him 24 stickers.
Jerry ended up with 139 stickers.
How many stickers did Jerry start with?

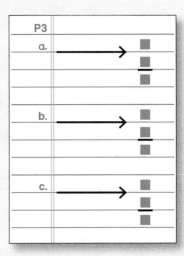

Part 4 Copy each problem and work it.

P4				
a.	803 − 2 5 1	b. 6 2 9 + 1 9 4	c. 4 5 7 + 9 3	d. 4 8 0 − 1 4 2

Part 5 Write the missing number for each fact.

a. ■ inches equals 1 foot

b. ■ cents equals 1 dime

c. ■ feet equals 1 yard

d. ■ cents equals 1 dollar

e. ■ minutes equals 1 hour

f. ■ seconds equals 1 minute

g. ■ days equals 1 week

h. ■ months equals 1 year

i. ■ hours equals 1 day

P5				
a.	b.	c.	d.	e.
f.	g.	h.	i.	

Lesson 35

Part 1 For each problem, make a number family with the letters for **Start** and **End**. Then work the problem and write the unit name.

a. The teacher had some stickers.
   She gave away 45 stickers.
   She ended up with 86 stickers.
   How many stickers did she start with?

b. Mr. Smith had some dollars.
   He spent 119 dollars.
   He ended up with 85 dollars.
   How many dollars did he start with?

c. Jada had 40 flowers.
   She gave 14 flowers to her mother.
   How many flowers did Jada end up with?

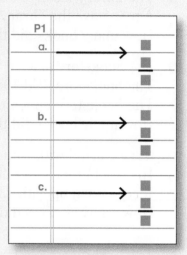

Part 2 Write **Cu** for cube, **Py** for pyramid, **Sp** for sphere, **H** for hexagon, **P** for pentagon, **R** for rectangle, and **T** for triangle.

 1.   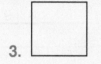 2.   3.   4.

 5.   6.   7.   8.

P2			
1.	2.	3.	4.
5.	6.	7.	8.

Part 3 Copy each problem and work it.

P3			
a. $4.96 − 1.90	b. $5.80 − 2.39	c. $4.90 − 1.25

Lesson 35

Part 4 Write the missing number for each item.

a. 132 131 ■ 129   b. 82 84 ■ 88

c. 55 50 ■ 40   d. 36 46 ■ 66

P4			
a.	b.	c.	d.

Part 5 Write the missing number for each fact.

a. ■ hours equals 1 day

b. ■ cents equals 1 quarter

c. ■ days equals 1 week

d. ■ inches equals 1 foot

e. ■ cents equals 1 dollar

f. ■ cents equals 1 dime

g. ■ minutes equals 1 hour

h. ■ months equals 1 year

i. ■ seconds equals 1 minute

j. ■ feet equals 1 yard

P5				
a.	b.	c.	d.	e.
f.	g.	h.	i.	j.

Lesson 36

Part 1

a. The truck went 24 more miles than the car went.

b. The boy bought 26 more apples.

c. Then the truck went 24 more miles.

d. The boy bought 26 more apples than his sister bought.

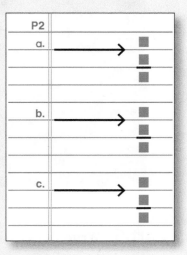

Independent Work

Part 2 Work each problem. Remember to write the unit name.

a. Jerry ate 78 more strawberries than Tom.
Jerry ate 109 strawberries.
How many strawberries did Tom eat?

b. Sally planted 269 trees.
Sally planted 131 fewer trees than José.
How many trees did José plant?

c. Kurt wrote 304 fewer letters than Alice.
Alice wrote 824 letters.
How many letters did Kurt write?

Connecting Math Concepts

Lesson 36

Part 3 | Write the fraction for each picture.

a.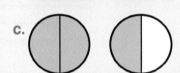

b.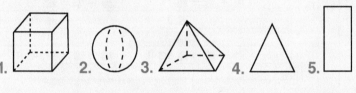

c.

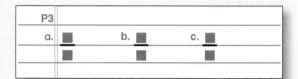

P3			
a. ■	b. ■	c. ■	
■	■	■	

Part 4 | Write **Cu** for cube, **Py** for pyramid, **Sp** for sphere, **H** for hexagon, **P** for pentagon, **R** for rectangle, **S** for square, **T** for triangle.

1. 2. 3. 4. 5.

P4				
1.	2.	3.	4.	5.
6.	7.	8.	9.	10.

6. 7. 8. 9. 10.

Part 5 | Copy and work each problem.

P5			
a.	3 5	b.	9 5
	1 8 5		8 0
	+ 2 5		+ 3 8

Lesson 37

Part 1

a. Fran ran 16 more miles.

b. The boy sold 26 fewer apples than his father sold.

c. The truck went 24 more miles than the car went.

d. Then the truck went 24 more miles.

e. She gave away 4 pies.

Part 2

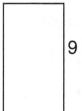

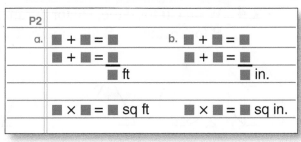

6 ft

2 ft

b. 4 in.

9 in.

Independent Work

Part 3 | Work each problem.

a. A boy had some flowers. He found 136 more flowers.
He ended up with 240 flowers.
How many flowers did the boy start out with?

b. There were 99 people in the park.
24 more people came into the park.
How many people ended up in the park?

c. The store had lots of apples. The store sold 79 apples.
At the end of the day, the store had 26 apples.
How many apples did the store have at the beginning of the day?

d. Alice had 149 pencils.
She gave away 79 pencils.
How many pencils did she have left?

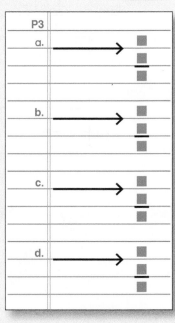

Connecting Math Concepts

Lesson 37

Part 4 Copy each problem and work it.

P4			
a. $9.40 − 2.25	b. $3.85 + 1.95	c. $24.20 +13.80	

Part 5 Write the fraction for each picture.

a.

b.

c.

P5			
a. ▪ ▪	b. ▪ ▪	c. ▪ ▪	

Part 6 Write the dollars and cents amount for each row.

a.

b.

P6	
a.	b.

Lesson 38

Part 1

a. Gary was 8 years younger than Tom.

b. The girl bought 9 more pencils than Jerry.

c. Heather found 36 more red stones.

d. The train went 45 more miles.

e. 76 men got off the train.

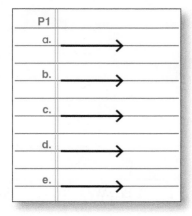

Part 2

	May	June	Total for both months
River City	15	5	20
Hill Town	10	30	40
Total for both cities	25	35	60

a. How many people visited Hill Town in June?

b. How many people visited both cities in May?

c. How many people visited River City in June?

d. How many people visited River City in both months?

Independent Work

Part 3

$1.95 $5.50 $ .45 $2.19

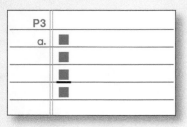

a. You want to buy the soap, the hairbrush, and the toothpaste. How much money do you need?

Lesson 38

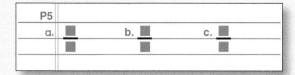

Part 4 Copy and work each regular problem. Next to the regular problem, write and work an estimation problem.

P4							
a.	4 2		b.	5 2		c.	6 3
	+ 3 6			+ 1 6			+ 3 7

Part 5 Write the fraction for each picture.

a.

b.

c.

P5			
a.	▪	b. ▪	c. ▪
	▪	▪	▪

Part 6 Work each problem. Remember to write the unit name.

a. Tim is 28 years younger than his dad.
   Tim is 43 years old.
   How old is his dad?

b. Fran walked 35 more miles than Al walked.
   Al walked 125 miles.
   How many miles did Fran walk?

c. Donna weighed 22 pounds less than Eric weighed.
   Eric weighed 150 pounds.
   How much did Donna weigh?

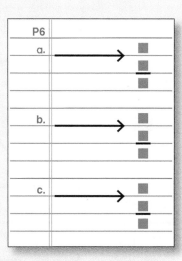

Lesson 39

Part 1

a. The green train had 400 more passengers than the red train.

b. There were 100 fewer brown bottles than yellow bottles.

c. Then the dog gained 16 pounds.

d. Then Nancy sold 13 pies.

e. The red car weighed 125 pounds less than the blue car.

f. The teacher gave away 15 stickers.

P1	
a.	⟶
b.	⟶
c.	⟶
d.	⟶
e.	⟶
f.	⟶

Part 2

a. 5 in. / 4 in. (square)

b. 2 ft / 5 ft (rectangle)

P2	
a. ■ + ■ = ■	b. ■ + ■ = ■
■ + ■ = ■ (underline)	■ + ■ = ■ (underline)
■	■
■ × ■ = ■	■ × ■ = ■

Part 3

	Clear Lake	Swan Lake	Total for both lakes
Small boats	15	22	37
Big boats	17	29	46
Total boats	32	51	83

P3					
a.	b.	c.	d.	e.	

a. How many small boats were on Swan Lake?

b. How many big boats were on both lakes?

c. How many big boats were on Clear Lake?

d. How many total boats were on Swan Lake?

e. How many small boats were on Clear Lake?

Lesson

Part 4

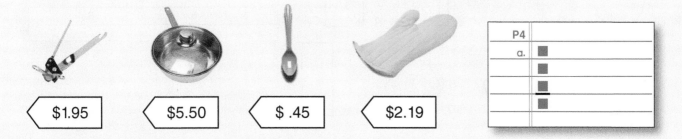

P4	
a.	■
	■
	■
	■

a. You want to buy the oven mitt, the pot, and the can opener. How much money do you need?

Part 5 Copy and work each problem.

P5			
a.	724 − 99	b. 530 −265	c. 492 −196

Part 6 Work each problem. Remember to write the unit name.

a. Henry had some rocks.
   He threw 24 rocks into the river.
   He ended up with 70 rocks.
   How many rocks did he start with?

b. Rita had 51 rocks.
   She found 37 more rocks.
   How many rocks did she end up with?

c. You have some marbles. You lose 37 marbles.
   You end up with 45 marbles.
   How many marbles did you start with?

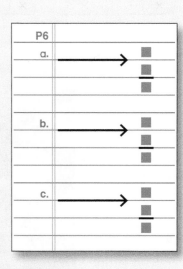

Lesson 39

Part 7 Write **Cu** for cube, **Py** for pyramid, **Sp** for sphere, **H** for hexagon, **P** for pentagon, **R** for rectangle, **S** for square, **T** for triangle.

 1.
 2.
 3.
 4.

P7			
1.	2.	3.	4.
5.	6.	7.	8.

 5.
 6.
 7.
 8.

Part 8 Write the fraction for each picture.

a.   b.

P8			
a. ▪	b. ▪	c. ▪	d. ▪
▪	▪	▪	▪

c. 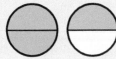   d.

Part 9 Write the answer to each problem.

a. $\begin{array}{r} 10 \\ \times\ 4 \\ \hline \end{array}$
b. $\begin{array}{r} 1 \\ \times 3 \\ \hline \end{array}$
c. $\begin{array}{r} 2 \\ \times 1 \\ \hline \end{array}$
d. $\begin{array}{r} 9 \\ \times 1 \\ \hline \end{array}$
e. $\begin{array}{r} 10 \\ \times\ 3 \\ \hline \end{array}$

P9				
a.	b.	c.	d.	e.

Connecting Math Concepts

Lesson 40

Part 1

a. John had 18 pencils.
   He gave away 5 of those pencils.
   How many pencils did he end up with?

b. The fir tree was 66 feet tall.
   The fir tree was 47 feet taller than the apple tree.
   How many feet tall was the apple tree?

c. Sally weighed 21 pounds more than Fred.
   Sally weighed 118 pounds.
   How much did Fred weigh?

d. The truck started out with 56 boxes.
   Then the truck picked up 147 more boxes.
   How many boxes did the truck end up with?

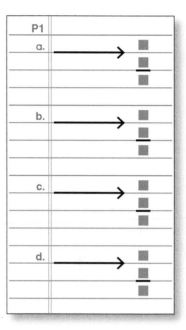

Independent Work

Part 2

a. Find the perimeter of this rectangle.

b. Find the area of this rectangle.

10 in.

3 in.

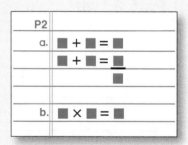

Part 3 Copy and work each regular problem. Next to the regular problem, write and work an estimation problem.

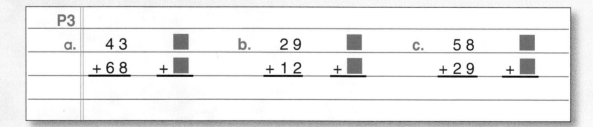

Lesson 40

Part 4 Copy and work each problem.

P4			
a.	370 −195	b. 752 −176	c. 640 −480

Part 5 Work each problem. Remember to write the unit name.

a. Alice was 15 inches taller than Maria.
   Maria was 47 inches tall.
   How many inches tall was Alice?

b. Tom was 27 inches shorter than Bob.
   Bob was 72 inches tall.
   How many inches tall was Tom?

c. Tyler was 50 pounds lighter than Greg.
   Tyler weighed 170 pounds.
   How many pounds did Greg weigh?

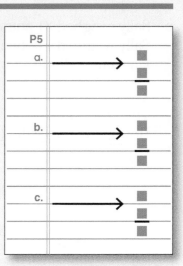

Part 6 Write the fraction for each picture.

a.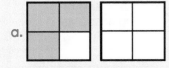

b.

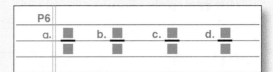

c.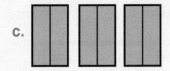

d.

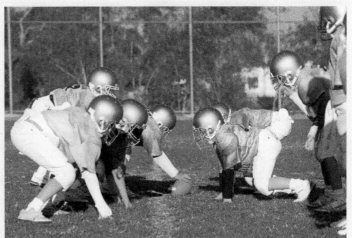

Lesson 41

Part 1

a. Tom had some stamps.
Then he found 18 more stamps.
He ended up with 60 stamps.
How many stamps did he start with?

b. The elm tree was 82 feet shorter than the pine tree.
The elm tree was 66 feet tall.
How tall was the pine tree?

c. Hilda's dad was 36 years older than Hilda.
Hilda was 19 years old.
How old was her dad?

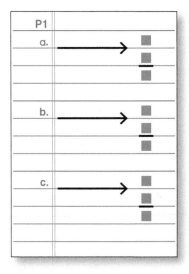

Part 2

	Monday	Tuesday	Total for both days
Red birds	21	78	99
Yellow birds	54	31	85
Total birds	75	109	184

P2				
a.		b.	c.	d.

a. How many yellow birds were seen on Monday?

b. How many total birds were seen on Monday?

c. How many red birds were seen on both days?

d. How many total birds were seen on Tuesday?

Independent Work

Part 3 Copy and work each problem.

P3				
a.	324 − 99	b. 305 −125	c. 820 −235	d. 742 − 39

Lesson 41

Part 4

a. Find the perimeter of this rectangle.

b. Find the area of this rectangle.

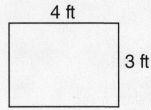

4 ft

3 ft

P4	
a.	■ + ■ = ■
	■ + ■ = ■
	■
b.	■ × ■ = ■

Part 5 | Write the missing number for each item.

a. 288 287 286 ■ 284

b. 155 160 165 ■ 175

c. 160 150 140 ■ 120

d. 36 46 56 ■ 76

P5			
a.	b.	c.	d.

Part 6 | Write **Cu** for cube, **Py** for pyramid, **Sp** for sphere, **H** for hexagon, **P** for pentagon, **R** for rectangle, **S** for square, **T** for triangle.

1.   2.   3. 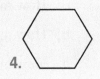 4.

P6			
1.	2.	3.	4.
5.	6.	7.	8.

5.  6.  7.  8.

Part 7

 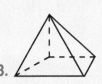

P7		
a.	■	b. ■
	■	■
	■	■
	■	■

$1.09 $2.93 $6.11 $2.06

a. A man buys bananas, meat, and milk. How much does he spend?

b. A boy buys bananas, juice, and meat. How much does he spend?

Connecting Math Concepts

Lesson

Part 8 Write the numbers for counting by 9s to 90.

P8	
a.	9 ■ ■ ■ ■ ■ ■ ■ ■ 90

Part 9 For each item, work the column problem to find the missing number.

a. 140 + ■ = 200   b. 315 + 299 = ■

P9			
a.	■	b.	■
	■		■
	■		■

Lesson 42

Part 1

a. A mail carrier had 113 letters in her bag.
   Then she picked up another 204 letters.
   How many letters did she have in all?

b. The yellow bottle weighed 36 pounds more than the blue bottle.
   The yellow bottle weighed 39 pounds.
   How many pounds did the blue bottle weigh?

c. The bulldog had 47 bones.
   The dog found 70 more bones.
   How many bones does it have now?

d. Tom had 57 fewer baseball cards than Frank.
   Frank had 180 baseball cards.
   How many cards did Tom have?

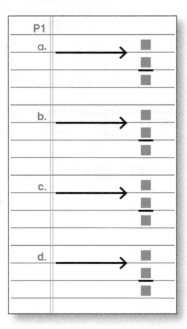

Part 2

	Mountain Park	Valley Park	Total for both parks
Small rocks	18	14	32
Large rocks	12	15	27
Total rocks	30	29	59

a. How many total rocks were there in Mountain Park?

b. How many small rocks were in Valley Park?

c. How many large rocks were there in both parks?

d. How many large rocks were in Mountain Park?

P2			
a.	b.	c.	d.

Independent Work

Part 3 Copy and work each problem.

P3			
a. $\begin{array}{r} 624 \\ -197 \end{array}$	b. $\begin{array}{r} 640 \\ -275 \end{array}$	c. $\begin{array}{r} 749 \\ -\ 64 \end{array}$	d. $\begin{array}{r} 520 \\ -490 \end{array}$

Connecting Math Concepts

Lesson 42

Part 4 | Write the numbers for counting by 4s to 40.

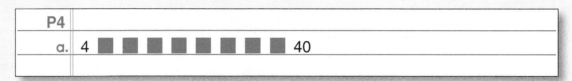

P4	
a.	4 ■ ■ ■ ■ ■ ■ ■ ■ 40

Part 5

a. Find the perimeter of this rectangle.

b. Find the area of this rectangle.

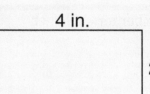

4 in.

2 in.

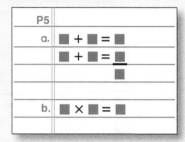

P5	
a.	■ + ■ = ■
	■ + ■ = ■
	■
b.	■ × ■ = ■

Part 6 | Write the fraction for each picture.

a.

b.

c.

d.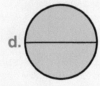

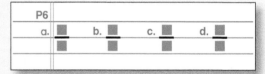

P6				
a. ■/■	b. ■/■	c. ■/■	d. ■/■

Part 7 | Copy and work each regular problem.
Next to the regular problem, write and work an estimation problem.

P7		
a.	7 9 ■	b. 9 2 ■
	– 4 2 – ■	– 4 3 – ■

Lesson 42

Part 8 Write the dollars and cents amount for each row.

a.

b.

P8	
a.	b.

Part 9 Write the missing number for each fact.

a. ▪ feet equals 1 yard

b. ▪ hours equals 1 day

c. ▪ seconds equals 1 minute

d. ▪ months equals 1 year

P9			
a.	b.	c.	d.

Connecting Math Concepts

Lesson

Part 1

a.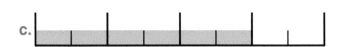

b.

c.

P1			
a. ■	b. ■	c. ■	
■	■	■	

Part 2

a. The truck started out with some boxes.
Workers put 27 more boxes on the truck.
The truck ended up with 49 boxes.
How many boxes did the truck start with?

b. The bed weighed 88 pounds more than the chair.
The chair weighed 19 pounds.
How many pounds did the bed weigh?

c. Tom's cat was 13 inches tall.
Tom was 43 inches taller than his cat.
How many inches tall was Tom?

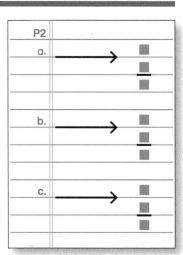

P2	
a. ⟶	■ ■ ■
b. ⟶	■ ■ ■
c. ⟶	■ ■ ■

Independent Work

Part 3 | Write each number.

a. six thousand two hundred

b. seven thousand six

c. nine thousand ninety

d. four thousand one

P3			
a.	b.	c.	d.

Part 4 | Copy each item and write the sign >, <, or =.

P4	
a. $\frac{7}{7}$ ■ 1 b. $\frac{9}{4}$ ■ 1 c. $\frac{9}{10}$ ■ 1 d. $\frac{5}{5}$ ■ 1 e. $\frac{3}{2}$ ■ 1

Lesson 43

Part 5 Copy and work each problem.

P5			
a.	434 −196	b. 230 − 98	c. 340 − 70

Part 6

a. Find the perimeter of this rectangle.

b. Find the area of this rectangle.

5 cm

2 cm

P6	
a.	■ + ■ = ■
	■ + ■ = ■
	■
b.	■ × ■ = ■

Part 7

$1.09 $2.33 $2.06 $3.00

P7		
a. ■		b. ■
■		■
■		■
■		■

a. Valerie buys juice, milk, and cereal. How much does she spend?

b. A man buys bananas, juice, and cereal. How much does he spend?

Part 8 Write the missing number for each fact.

a. ■ feet equals 1 yard
b. ■ inches equals 1 foot
c. ■ months equals 1 year
d. ■ seconds equals 1 minute
e. ■ hours equals 1 day

P8				
a.	b.	c.	d.	e.

Lesson 44

Part 1

	Fall	Winter	Total for both seasons
Rainbow Valley	106	54	160
Big Lake	206	56	262
Total for both places	312	110	422

P1				
a.	b.	c.	d.	e.

a. How many people visited Big Lake in both seasons?

b. How many people visited Rainbow Valley in the fall?

c. How many people visited both places in the fall?

d. How many people visited Rainbow Valley in both seasons?

e. In the winter, did more vistors go to Rainbow Valley or Big Lake?

Part 2

a. $8.26   b. $7.09   c. $14.90

d. $8.29   e. $6.81

P2				
a.	b.	c.	d.	e.

Part 3

a.

b.

c.

d.

e.

f.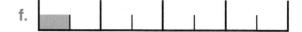

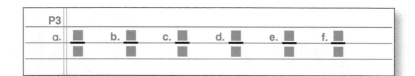

P3						
a.	b.	c.	d.	e.	f.

Lesson

Part 4 Find the sentence that tells how to make the number family. Make the number family for that sentence and then work the problem. Remember to write the unit name.

a. The chair was 51 years old.
   The lamp was 37 years older than the chair.
   How old was the lamp?

b. The bear weighed 280 pounds in the spring.
   The bear gained 227 pounds.
   How many pounds did the bear end up weighing?

c. The store had 156 cans of soup.
   During the day, the store sold 47 cans of soup.
   How many cans of soup did the store end up with?

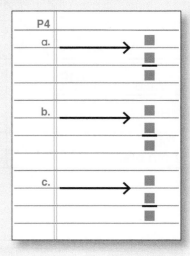

Part 5 Write each number.

a. nine thousand nine hundred nine

b. six thousand four

c. three thousand five hundred seven

d. one thousand one hundred thirty-nine

e. eight thousand fourteen

P5				
a.	b.	c.	d.	e.

Lesson 44

Part 6 Copy each item and write the sign >, <, or =.

P6					
a. $\dfrac{4}{3}$ ■ 1	b. $\dfrac{7}{9}$ ■ 1	c. $\dfrac{2}{2}$ ■ 1	d. $\dfrac{3}{5}$ ■ 1	e. $\dfrac{5}{3}$ ■ 1

Part 7 Copy and work each problem.

P7			
a.	$\begin{array}{r} 420 \\ -389 \\ \hline \end{array}$	b.	$\begin{array}{r} 154 \\ -148 \\ \hline \end{array}$	c.	$\begin{array}{r} 620 \\ -\ 70 \\ \hline \end{array}$

Part 8 Copy and work each regular problem.
Next to the regular problem, write and work an estimation problem.

P8			
a.	$\begin{array}{r} 68 \\ +23 \\ \hline \end{array}$ ■ $\begin{array}{r} \\ +\ ■ \\ \hline \end{array}$	b.	$\begin{array}{r} 79 \\ -41 \\ \hline \end{array}$ ■ $\begin{array}{r} \\ -\ ■ \\ \hline \end{array}$	c.	$\begin{array}{r} 58 \\ +28 \\ \hline \end{array}$ ■ $\begin{array}{r} \\ +\ ■ \\ \hline \end{array}$

Part 9 Write the fraction for each picture.

a.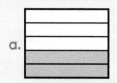

b.

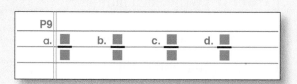

P9			
a. $\dfrac{■}{■}$	b. $\dfrac{■}{■}$	c. $\dfrac{■}{■}$	d. $\dfrac{■}{■}$

c.   d.

Lesson 45

Part 1

a.    b.

c.    d.

e.    f.

P1						
a. ▪	b. ▪	c. ▪	d. ▪	e. ▪	f. ▪	
▪	▪	▪	▪	▪	▪	

Part 2

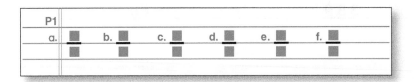

P2				
a.	$ 5.26	b. $ 2.78	c. $ 9.34	d. $ 5.79
	+ 3.71	− 1.98	+ 2.64	− 4.56
	▬▬▬	▬▬▬	▬▬▬	▬▬▬

Independent Work

Part 3 Write each number.

a. two thousand three hundred six

b. three thousand eight

c. three thousand five hundred seventy

d. two thousand six hundred forty-nine

e. five thousand eighteen

P3				
a.	b.	c.	d.	e.

Lesson 45

Part 4 Copy each item and write the sign >, <, or =.

P4	
a. $\frac{2}{5}$ ■ 1 b. $\frac{4}{4}$ ■ 1 c. $\frac{3}{2}$ ■ 1 d. $\frac{8}{8}$ ■ 1	

Part 5 Find the sentence that tells how to make the number family. Make the family for that sentence and then work the problem. Remember to write the unit name.

a. Nancy had some books. She gave 95 books away.
   She ended up with 69 books.
   How many books did Nancy start with?

b. Tim had 527 stamps.
   Joe had 319 fewer stamps than Tim had.
   How many stamps did Joe have?

c. A raccoon picked up 185 more sticks than a beaver picked up.
   The beaver picked up 246 sticks.
   How many sticks did the raccoon pick up?

P5	
a. ⟶ ■ ■ ■	
b. ⟶ ■ ■ ■	
c. ⟶ ■ ■ ■	

Part 6 Write **Cu** for cube, **Py** for pyramid, **Sp** for sphere, **H** for hexagon, **P** for pentagon, **R** for rectangle, **S** for square, and **T** for triangle.

1.   2.   3.   4.

P6			
1.	2.	3.	4.
5.	6.	7.	8.

5.   6.   7.   8.

Part 7

a. Find the perimeter of this rectangle.

b. Find the area of this rectangle.

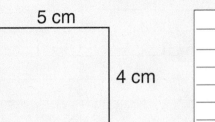

5 cm

4 cm

P7	
a.	■ + ■ = ■
	■ + ■ = ■
	■
b.	■ × ■ = ■

Lesson 46

Part 1

a.
| dodgeball |
| game |
| tag |

b.
| pear |
| apple |
| fruit |

c.
| house |
| building |
| garage |

d.
| boy |
| children |
| girl |

e.
| bug |
| spider |
| ant |

P1	
a.	→
b.	→
c.	→
d.	→
e.	→

Part 2

a.
$$\$ 4.32$$
$$+ 3.91$$

b.
$$\$ 6.89$$
$$- 3.43$$

c.
$$\$ 8.48$$
$$+ 1.41$$

d.
$$\$ 5.63$$
$$+ 5.23$$

P2				
a. ■	b. ■	c. ■	d. ■	
■	■	■	■	
■	■	■	■	

Part 3

a.

d.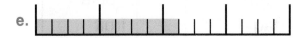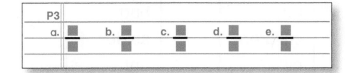

b.

e.

c.

P3					
a. ■	b. ■	c. ■	d. ■	e. ■
■	■	■	■	■

Part 4

a. How much more is 85 than 61?

b. How much less is 217 than 348?

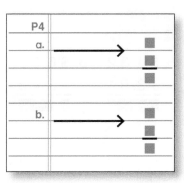

Independent Work

Part 5 Work each problem. Remember to write the unit name.

a. The store had 634 bottles.
   The store sold 144 bottles.
   How many bottles did the store end up with?

b. Jenny started out with some cans.
   She found 36 more cans. She ended up with 58 cans.
   How many cans did she start with?

c. Brian had 400 fewer apples than Ginger had.
   Ginger had 599 apples.
   How many apples did Brian have?

d. A horse weighed 120 pounds less than a cow.
   The horse weighed 730 pounds.
   How much did the cow weigh?

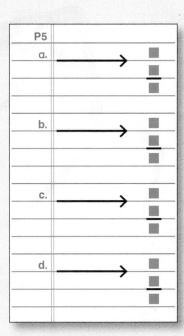

Lesson 46

Part 6 Write each number.

a. five thousand

b. five hundred

c. two thousand five hundred

d. one thousand one hundred eleven

e. one thousand twenty

P6				
a.	b.	c.	d.	e.

Part 7 Copy and work each problem.

P7				
a.	532 − 96	b. 285 +137	c. 950 − 72	d. 560 −190

Part 8 Write **Cu** for cube, **Py** for pyramid, **Sp** for sphere, **H** for hexagon, **P** for pentagon, **R** for rectangle, **S** for square, **T** for triangle.

1. 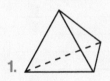 2. 3. 4.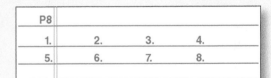

P8			
1.	2.	3.	4.
5.	6.	7.	8.

5. 6. 7. 8.

Part 9 Write the fraction for each picture.

a. b.

P9			
a.	b.	c.	d.

c. d.

Lesson 47

Part 1

a.
```
tiger
monkey
animal
```

b.
```
hamburger
food
cereal
```

c.
```
vehicle
car
truck
```

d.
```
saw
tool
drill
```

e.
```
brother
uncle
relative
```

P1

a. →

b. →

c. →

d. →

e. →

Part 2

a. How much less is 12 than 47?

b. How much more is 86 than 49?

c. How much more is 176 than 56?

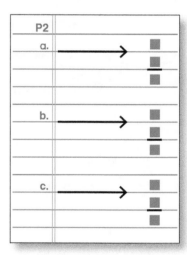

Part 3

P3

a.
```
$ 1.9 0     $ ■■
+ 8.9 6     + ■■
```

b.
```
$ 8.6 7     $ ■■
+ 4.3 2     + ■■
```

c.
```
$ 6.6 8     $ ■■
+ 2.2 5     + ■■
```

Lesson

Part 4 Write each number.

a. one thousand sixty

b. two thousand one hundred

c. three thousand five hundred

d. five thousand six hundred thirty-nine

e. one thousand six

P4				
a.	b.	c.	d.	e.

Part 5 Work each problem. Remember to write the unit name.

a. Kathy picked 47 more apples than Debby.
   Kathy picked 249 apples.
   How many apples did Debby pick?

b. A dog started out with some fleas.
   Then the dog got rid of 48 fleas.
   The dog ended up with 99 fleas.
   How many fleas did the dog start with?

c. A tree was 51 feet tall.
   The tree was 19 feet taller than the building.
   How many feet tall was the building?

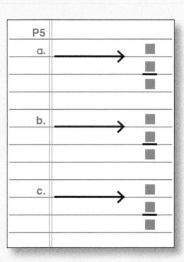

Part 6 Copy and work each problem.

P6			
a.	5 2 4 0	b. 3 2 0	c. 4 2 5
	+ 9 9 9	− 1 9 9	− 1 6 9

Lesson

Part 7 Write the missing number for each fact.

a. ▦ inches equals 1 foot

b. ▦ cents equals 1 dime

c. ▦ feet equals 1 yard

d. ▦ cents equals 1 dollar

e. ▦ minutes equals 1 hour

f. ▦ seconds equals 1 minute

g. ▦ days equals 1 week

h. ▦ months equals 1 year

i. ▦ hours equals 1 day

P7				
a.	b.	c.	d.	e.
f.	g.	h.	i.	

Part 8

a. Find the perimeter of this rectangle.

b. Find the area of this rectangle.

5 cm

2 cm

P8	
a.	▦ + ▦ = ▦
	▦ + ▦ = ▦
	▦
b.	▦ × ▦ = ▦

Lesson

P1				
a.	$ 5.6 2 + 3.2 5	$ ▪▪ + ▪▪	b. $ 4.0 8 + 5.9 8	$ ▪▪ + ▪▪
c.	$ 8.7 6 − 5.4 3	$ ▪▪ − ▪▪		

Part 2

a. How much older is the red car than the blue car?

b. How much shorter is Sam than Bob?

c. How much taller is the house than the fence?

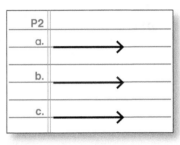

Independent Work

Part 3 Work each problem. Remember to write the unit name.

a. A cow weighed 196 pounds less than a horse.
   The cow weighed 741 pounds.
   How many pounds did the horse weigh?

b. A truck was 37 inches longer than a car.
   The truck was 219 inches long.
   How many inches long was the car?

c. A toy store had some dolls.
   Then the store gave away 124 dolls.
   The store ended up with 525 dolls.
   How many dolls did the store start with?

d. There were 94 passengers on the bus.
   27 more passengers got on the bus.
   How many passengers did the bus end up with?

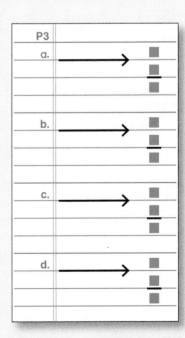

Lesson 48

Part 4 Copy each item and write the sign >, <, or =.

P4			
a.	12 ■ 7 + 7	b. 20 – 10 ■ 10	c. 5 + 3 ■ 20

Part 5 Copy and work each problem.

P5			
a.	324 – 99	b. 305 –125	c. 820 –235

Part 6

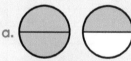
$1.05

$2.45

$ .80

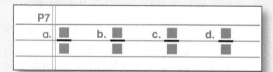
$2.99

P6		
a. ■		b. ■
■		■
■		■
■		

a. Alicia buys bananas, milk, and cereal. How much does she spend?

b. A man buys milk and juice. How much does he spend?

Part 7 Write the fraction for each picture.

a.

b.

P7			
a. ■	b. ■	c. ■	d. ■
■	■	■	■

c.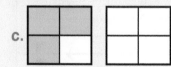

d.

Part 8 Copy and work each regular problem.
Next to the regular problem, write and work an estimation problem.

P8					
a.	8 1	■	b.	5 2	■
	− 4 9	− ■		+ 3 8	+ ■

Lesson 49

Part 1

a. Tom is 14 years old.
Mary is 39 years old.
How many years older is Mary than Tom?

b. The car went 420 miles.
The truck went 720 miles.
How many more miles did the truck go than the car?

c. Jan is 66 inches tall.
Tim is 71 inches tall.
How much taller is Tim than Jan?

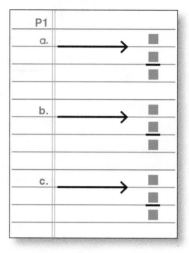

Part 2

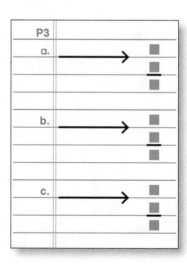

a.
$4.04
+5.94

$ ▢▢
+ ▢▢

b.
$8.82
−4.91

$ ▢▢
− ▢▢

Part 3

a. There are 32 children at the park.
14 of them are girls.
How many of them are boys?

b. There were 81 animals in the field.
15 were dogs. The rest were cats.
How many cats were in the field?

c. There were 19 long pencils
and 14 short pencils
How many pencils were there altogether?

Connecting Math Concepts

Lesson 49

Independent Work

Part 4 Copy each item and write the sign >, <, or =.

P4	
a.	4 + 5 ■ 5 + 5 b. 5 × 2 ■ 10 − 1 c. 5 × 3 ■ 10 + 5

Part 5 Work each problem. Remember to write the unit name.

a. The river is 184 yards longer than the lake.
   The lake is 59 yards long.
   How many yards long is the river?

b. Sam had 430 nails.
   He used 150 of the nails.
   How many nails did he end up with?

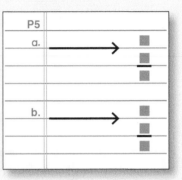

Lesson 49

Part 6 Copy and work each problem.

P6			
a.	7410	b. 7054	c. 840
	−3109	−1200	− 99

Part 7 Copy and work each regular problem.
Next to the regular problem, write and work an estimation problem.

P7		
a.	42 ■	b. 18 ■
	+37 +■	+43 +■

Part 8 Write the missing number for each item.

a. 185 184 183 ■ 181

b. 140 130 120 ■ 100

c. 862 863 864 ■ 866

d. 155 160 165 ■ 175

e. 36 46 56 ■ 76

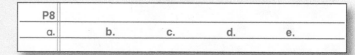

P8				
a.	b.	c.	d.	e.

Part 9 Work each problem.

a. How much less is 485 than 925?

b. How much more is 320 than 150?

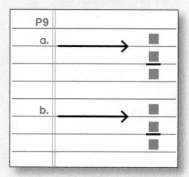

Lesson

a. John is 21 years old.
   Dan is 35 years old.
   How many years older is Dan than John?

b. Marsha weighs 85 pounds.
   Ken weighs 196 pounds.
   How many more pounds does Ken weigh than Marsha?

c. The truck is 34 feet long.
   The car is 9 feet long.
   How much longer is the truck than the car?

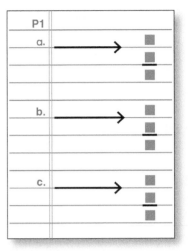

Part 2

a. There were 14 cars.
   There were 66 trucks.
   How many vehicles were there?

b. There were apples and pears.
   There were 18 pieces of fruit in all.
   There were 12 pears.
   How many apples were there?

c. There were 32 plates in the kitchen.
   15 of the plates were dirty.
   How many plates were clean?

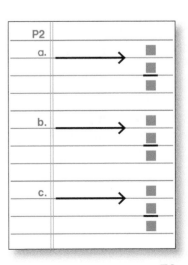

Part 3

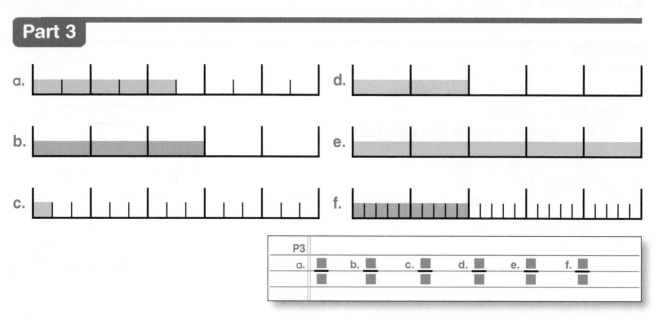

Independent Work

Part 4 Write each number.

a. one thousand sixteen

b. five thousand six hundred

c. three thousand nine

d. one thousand one hundred twenty

e. five thousand two

P4					
a.	b.	c.	d.	e.	

Part 5 Work each problem. Remember to write the unit name.

a. Sally planted 131 fewer trees than José.
Sally planted 269 trees.
How many trees did José plant?

b. Jerry had some marbles.
He found 41 marbles.
He ended up with 325 marbles.
How many marbles did he start with?

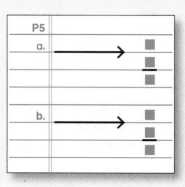

Lesson 50

Part 6 | Copy each item and write the sign >, <, or =.

	P6	
a.	9×2 ▇ $10 + 6$ b. $5 - 1$ ▇ $2 + 2$ c. $9 - 1$ ▇ 5×2	

Part 7 | Copy and work each problem.

	P7	
a.	$\begin{array}{r} 2\,6\,4\,0 \\ -\,1\,1\,8\,8 \\ \hline \end{array}$ b. $\begin{array}{r} 3\,8\,2\,5 \\ -\,1\,7\,7\,5 \\ \hline \end{array}$ c. $\begin{array}{r} 7\,4\,2 \\ +\ \ 9\,9 \\ \hline \end{array}$	

Part 8

a. Find the perimeter of this rectangle.

b. Find the area of this rectangle.

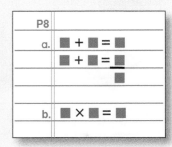

4 ft

2 ft

Part 9 | Work each problem.

a. How much more is 327 than 295?

b. How much less is 1540 than 2170?

Lesson 51

Part 1

a. There are 36 children in the park.
15 of them are boys.
How many are girls?

b. There are 19 hammers and 37 wrenches in a bag.
How many tools are there in the bag?

c. There are 254 apples in the store.
149 of the apples are red.
The rest of the apples are green.
How many green apples are in the store?

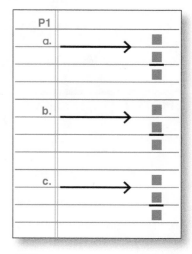

Part 2

$1.87 $9.26 $2.43 $4.85 $2.70

a. A woman buys the nuts and the hat.
About how much does she spend?

b. A man buys the book and the pen.
About how much does he spend?

c. A girl buys nuts, the hot dog, and the pen.
About how much does she spend?

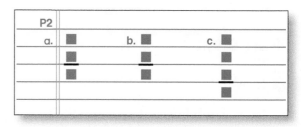

Lesson 51

Part 3

a. June rode a bike 156 miles farther than Ginger rode.
Ginger rode 97 miles.
How many miles did June ride?

b. Mike ate 123 apples.
Fran ate 86 apples.
How many more apples did Mike eat than Fran?

c. Sal weighs 137 pounds.
Jan weighs 109 pounds.
How many pounds heavier is Sal than Jan?

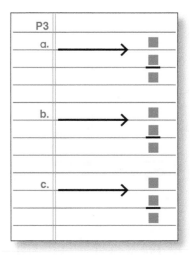

Part 4

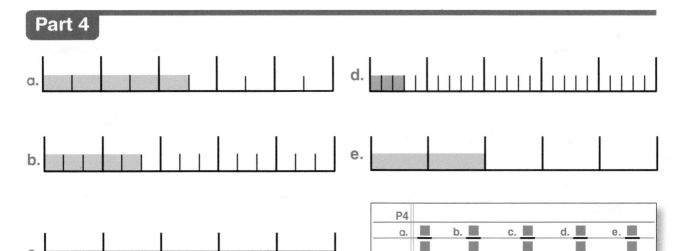

Independent Work

Part 5 Write each number.

a. four hundred

b. four thousand

c. nine thousand sixty

d. two thousand forty

e. six thousand six

Lesson 51

Part 6 Copy and work each problem.

P6			
a.	3 2 0 0 − 9 9 0	b. 4 2 0 − 9 9	c. 2 0 2 0 −1 9 2 0

Part 7 Write the fraction for each picture.

a.   b.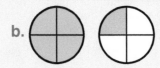

P7		
a. ▪/▪	b. ▪/▪	

Part 8 Copy each item and write the sign >, <, or =.

P8			
a. 9 − 1 ▪ 5 × 2	b. 8 + 7 ▪ 5 × 3	c. 10 + 10 ▪ 8 + 6	

Part 9

a. Find the perimeter of this rectangle.

b. Find the area of this rectangle.

10 cm

4 cm

P9	
a.	▪ + ▪ = ▪ ▪ + ▪ = ▪ ▪
b.	▪ × ▪ = ▪

Part 10 Copy and work each regular problem.
Next to the regular problem, work an estimation problem.

P10			
a.	9 1 − 5 9	▪ − ▪	b. 4 8 +2 3	▪ + ▪

Lesson 51

Part 11 Write the answer to each problem.

a. $5 \overline{)20}$ b. $5 \overline{)10}$ c. $5 \overline{)5}$

d. $5 \overline{)15}$ e. $5 \overline{)25}$

P11					
a.		b.	c.	d.	e.

Part 12 Write the numbers for counting by 4s to 40 and 9s to 90.

P12	
a.	4 ■ ■ ■ ■ ■ ■ ■ ■ 40
b.	9 ■ ■ ■ ■ ■ ■ ■ ■ 90

Lesson 52

Part 1

a. 25 + 3 = ◼

b. 74 + 4 = ◼

c. 83 + 6 = ◼

d. 44 + 5 = ◼

P1				
a. ◼	b. ◼	c. ◼	d. ◼	

Part 2

| $1.09 | $2.93 | $6.11 | $2.06 | $3.88 |

a. Valerie buys juice, milk, and cereal.
About how much does she spend?

b. A man buys bananas, meat, and milk.
Exactly how much does he spend?

c. A boy buys bananas, juice, and meat.
About how much does he spend?

d. A boy buys bananas, juice, and meat.
Exactly how much does he spend?

P2				
a. ◼	b. ◼	c. ◼	d. ◼
◼	◼	◼	◼
◼	◼	◼	◼
◼	◼	◼	◼

Part 3

a. The green truck was 39 feet long.
The red truck was 16 feet long.
How much longer was the green truck than the red truck?

b. Carla was 17 centimeters taller than her brother.
Her brother was 132 centimeters.
How tall was Carla?

c. John had 146 dollars.
Donna had 192 dollars.
How many more dollars did Donna have than John?

d. The car was 16 feet long.
The train was 265 feet longer than the car.
How long was the train?

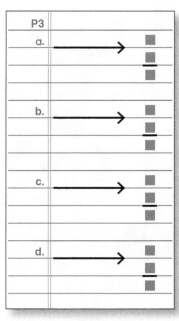

Connecting Math Concepts

Lesson

Independent Work

Part 4 | Work each problem. Write the unit name.

a. There are 457 children in the park.
275 of the children are boys.
How many girls are in the park?

b. The train started out with 825 passengers.
75 passengers got off the train.
How many passengers ended up on the train?

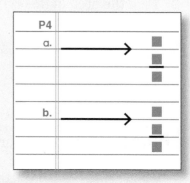

Part 5 | Copy each item and write the sign >, <, or =.

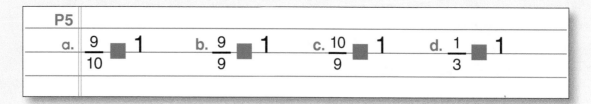

P5

a. $\frac{9}{10}$ ■ 1   b. $\frac{9}{9}$ ■ 1   c. $\frac{10}{9}$ ■ 1   d. $\frac{1}{3}$ ■ 1

Part 6 | Work each problem.

a. How much more is 2450 than 850?

b. How much less is 430 than 900?

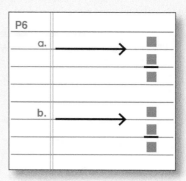

Part 7 | Write the missing number for each fact.

a. ■ seconds equals 1 minute   e. ■ inches equals 1 foot

b. ■ days equals 1 week   f. ■ minutes equals 1 hour

c. ■ months equals 1 year   g. ■ feet equals 1 yard

d. ■ hours equals 1 day   h. ■ cents equals 1 dollar

P7			
a.	b.	c.	d.
e.	f.	g.	h.

Connecting Math Concepts

Lesson 52   **87**

Lesson 52

Part 8

a. Find the perimeter of this rectangle.

b. Find the area of this rectangle.

4 ft

2 ft

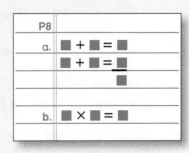

P8	
a.	■ + ■ = ■
	■ + ■ = ■
	■
b.	■ × ■ = ■

Part 9 Write the dollars and cents amount for each row.

a.

b.

c.

P9			
a.		b.	c.

Part 10 Write the answer to each problem.

a. $5\overline{)10}$ b. $5\overline{)20}$ c. $5\overline{)5}$

d. $5\overline{)25}$ e. $5\overline{)15}$

P10				
a.	b.	c.	d.	e.

Part 11 Write the numbers for counting by 4s to 40 and 9s to 90.

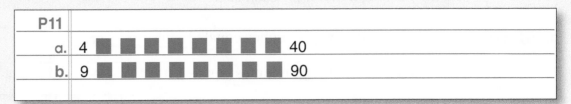

P11	
a.	4 ■ ■ ■ ■ ■ ■ ■ ■ 40
b.	9 ■ ■ ■ ■ ■ ■ ■ ■ 90

Lesson 53

Part 1

a. The brown house was 43 feet wide.
The yellow house was 65 feet wide.
How much wider was the yellow house than the brown house?

b. Pete was 8 years older than his brother.
Pete was 45 years old.
How old was his brother?

c. Sam had 16 pencils.
Jill had 23 pencils.
How many more pencils did Jill have than Sam?

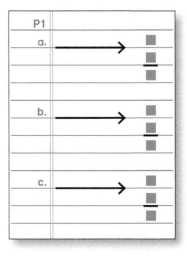

Part 2

 $10.99 $3.05 $5.02 $ .95 $4.01

a. A boy bought a notebook, paper, and a calculator.
Exactly how much did he spend?

b. A girl bought a notebook, paper, and a stapler.
About how much did she spend?

c. Susie bought a calculator, a stapler, and staples.
Exactly how much did she spend?

d. Susie bought paper and a notebook.
About how much did she spend?

Part 3

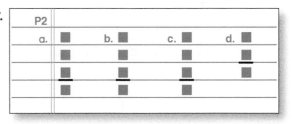

a. 55 + 3 = ▇   b. 27 + 2 = ▇   c. 82 + 6 = ▇

d. 94 + 4 = ▇   e. 23 + 2 = ▇   f. 41 + 7 = ▇

Lesson 53

Part 4 Copy and work each problem.

P4		
a.	$ 1 3 . 9 5	b. $ 2 4 . 3 0
	+ 6 . 9 0	+ 1 9 . 7 0

Part 5 Work each problem. Write the unit name.

a. There are 125 boys and 96 girls in the park.
   How many children are in the park?

b. The store started out with lots of stamps.
   The store sold 195 stamps.
   The store ended up with 485 stamps.
   How many stamps did the store start with?

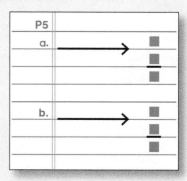

Part 6 Copy each item and write the sign >, <, or =.

P6		
a.	9 − 1 ▇ 4 + 5	b. 2 × 3 ▇ 2 + 3

Part 7 Write **Cu** for cube, **Py** for pyramid, **Sp** for sphere, **H** for hexagon,
P for pentagon, **R** for rectangle, **S** for square, **T** for triangle.

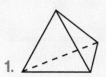
1.

2.

3.

4.

P7			
1.	2.	3.	4.
5.	6.	7.	8.

5.

6.

7.

8.

Lesson 53

Part 8 Copy and work each regular problem.
Next to the regular problem, work an estimation problem.

P8			
a.	6 2 + 2 9	■ + ■	b. 4 8 + 3 1	■ + ■

Part 9 Copy and work each problem.

P9		
a.	5 0 2 0 − 1 3 1 9	b. 4 5 3 0 − 9 9 0

Part 10 Copy each item and write the sign >, <, or =.

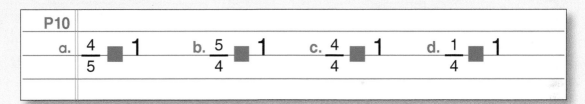

P10				
a. $\frac{4}{5}$ ■ 1	b. $\frac{5}{4}$ ■ 1	c. $\frac{4}{4}$ ■ 1	d. $\frac{1}{4}$ ■ 1	

Part 11 Write the answer to each problem.

a. $5\overline{)15}$ b. $5\overline{)25}$ c. $5\overline{)10}$

P11				
a.	b.	c.	d.	e.

d. $5\overline{)15}$ e. $5\overline{)20}$

Part 12 Write the numbers for counting by 4s to 40 and 9s to 90.

P12	
a.	4 ■ ■ ■ ■ ■ ■ ■ 40
b.	9 ■ ■ ■ ■ ■ ■ ■ 90

Lesson 54

Part 1

a. $\blacksquare + 29 = 61$ b. $35 + \blacksquare = 90$

c. $\blacksquare + 24 = 59$ d. $22 + \blacksquare = 49$

P1				
a. $\blacksquare$	b. $\blacksquare$	c. $\blacksquare$	d. $\blacksquare$
$\blacksquare$	$\blacksquare$	$\blacksquare$	$\blacksquare$
$\blacksquare$	$\blacksquare$	$\blacksquare$	$\blacksquare$

Part 2

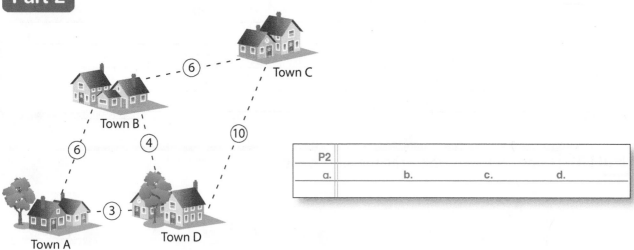

P2				
a.	b.	c.	d.

a. Meg started in Town D. She went to Town C. She did not go through any other towns.

b. Peg started in Town D. She went to Town C, but she went through one other town.

c. Ben starts in Town B and goes to Town C. He does not go through any other town.

d. Hank starts in Town B and goes to Town C, but he goes through one other town.

Part 3

a. John had $11.25.
   Then he earned $12.35.
   How much did he end up with?

b. Fran had $34.75.
   She bought shoes that cost $27.50.
   How much money did she end up with?

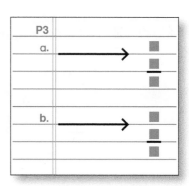

Lesson

Part 4

  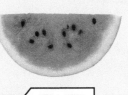  

| $3.89 | $3.07 | $4.92 | $4.05 | $5.11 |

a. A man bought bananas, apples, and a watermelon.
   Exactly how much did he spend?

b. A girl bought bananas, apples, and a watermelon.
   About how much did she spend?

c. Susie bought grapes and a pineapple.
   Exactly how much did she spend?

d. Gary bought grapes and a pineapple.
   About how much did he spend?

P4			
a. ■	b. ■	c. ■	d. ■
■	■	■	■
■	■	■	■
■	■		

Part 5 | Copy and work each problem.

P5					
a.	3 2 0 0	b.	4 2 4	c.	3 2 0 0
	− 1 1 2 0		− 9 9		− 9 9 0

Part 6 | Work each problem. Remember to write the unit name.

a. There are 75 clean plates and 95 dirty plates in the kitchen.
   How many plates are in the kitchen?

b. A plane travelled 2420 miles.
   A bus travelled 800 miles.
   How many more miles did the plane travel than the bus?

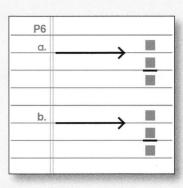

Lesson

Part 7 Write the fraction for each picture.

a.   b.

P7			
a. ▪		b. ▪	
—		—	
▪		▪	

Part 8 Write each number.

a. nine thousand

b. nine hundred

c. three thousand seven hundred

d. two thousand twelve

e. three thousand two hundred

P8					
a.		b.	c.	d.	e.

Part 9 Work each problem.

a. How much less is 350 than 800?

b. How much more is 1500 than 250?

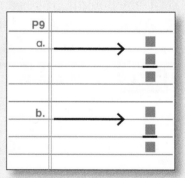

Part 10 Write the answer to each problem.

a. 5 ×3	b. 2 ×4	c. 3 ×2	d. 5 ×5	e. 2 ×1

P10					
a.	b.	c.	d.	e.
f.	g.	h.	i.	j.

f. 2 ×5	g. 4 ×5	h. 2 ×2	i. 1 ×5	j. 4 ×2

Connecting Math Concepts

Lesson 55

Part 1

P1				
a. ■ : ■	b. ■ : ■	c. ■ : ■	d. ■ : ■	

a.    b.    c.   d.

Part 2

a. $6.87   b. $1.05   c. $3.10   d. $4.94   e. $5.92

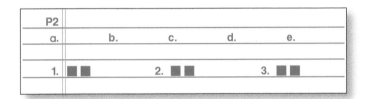

1. Jamie bought two of these items.
   One item was item A.
   He spent about 13 dollars.
   Which two items did he buy?

P2				
a.	b.	c.	d.	e.
1. ■ ■		2. ■ ■		3. ■ ■

2. Ann bought two items.
   One item was item D.
   She spent about 11 dollars.
   Which two items did she buy?

3. Ms. Bell bought two items.
   One item was item B.
   She spent about 4 dollars.
   Which two items did she buy?

Part 3

a. ■ + 14 = 37   b. 32 + ■ = 97   c. ■ + 14 = 88

P3			
a. ■	b. ■	c. ■	
■	■	■	
■	■	■	

Lesson 55

Part 4

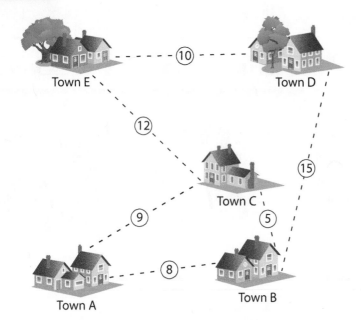

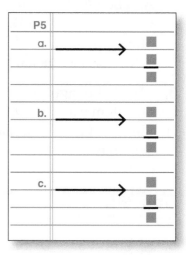

a. How many miles is the shortest route from Town B to Town A?

b. How many miles is the longest route from Town B to Town A?

c. How many miles is the route from Town B to Town A that goes through one other town?

Part 5

a. The window costs $21.00 more than the door.
The door costs $88.99.
How much does the window cost?

b. Tom had $44.68.
He bought a book that cost $18.35.
How much money did he still have?

c. The chicken dinner costs $9.20.
The burger dinner costs $7.20.
How much more does the chicken dinner cost
than the burger dinner?

Lesson

Part 6

  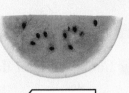  

| $3.89 | $3.07 | $4.92 | $4.05 | $5.11 |

a. A man bought apples, grapes, and a pineapple.
   Exactly how much did he spend?

b. A girl bought apples, grapes, and a pineapple.
   About how much did she spend?

c. Susie bought bananas and a watermelon.
   Exactly how much did she spend?

d. Gary bought bananas and a watermelon.
   About how much did he spend?

P6				
a. ▪	b. ▪	c. ▪	d. ▪	
▪	▪	▪̲	▪̲	
▪	▪			
▪̲	▪̲			

Part 7 | Copy and work each problem.

P7			
a.	5 4 0 0	b. 4 0 2 0	c. 3 3 3
	− 7 5 0	− 5 0 5	− 8 8

Part 8 | Work each problem.

a. How much more is 1500 than 700?

b. How much less is 3100 than 5000?

P8	
a. ⟶	▪
	▪
	▪̲
b. ⟶	▪
	▪
	▪

Lesson 55

Part 9 Write the missing number for each fact.

a. ☐ inches equals 1 foot

b. ☐ feet equals 1 yard

c. ☐ minutes equals 1 hour

d. ☐ cents equals 1 quarter

e. ☐ days equals 1 week

f. ☐ seconds equals 1 minute

g. ☐ months equals 1 year

h. ☐ hours equals 1 day

P9				
a.	b.	c.	d.	
e.	f.	g.	h.	

Part 10 Write the answer to each problem.

a. 1 × 5   b. 2 × 2   c. 4 × 5   d. 2 × 5   e. 2 × 1

P10				
a.	b.	c.	d.	e.
f.	g.	h.	i.	j.

f. 5 × 5   g. 3 × 2   h. 2 × 4   i. 5 × 3   j. 4 × 2

Part 11 Write the numbers for counting by 4s to 40 and 9s to 90.

P11	
a.	4 ☐ ☐ ☐ ☐ ☐ ☐ ☐ 40
b.	9 ☐ ☐ ☐ ☐ ☐ ☐ ☐ 90

Lesson 56

Part 1

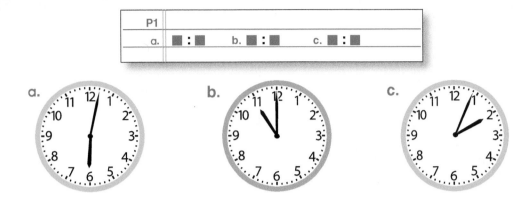

a.
b.
c.

Part 2

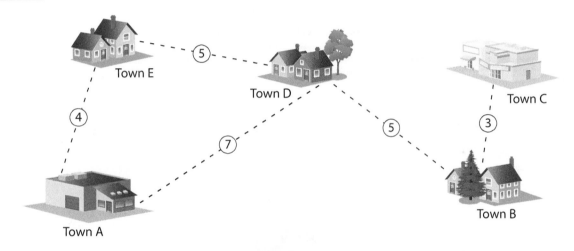

Town E

Town D

Town C

Town A

Town B

1. How many miles is the shortest route from Town A to Town B?

2. How many miles is the longest route from Town A to Town B?

3. How many miles is the shortest route from Town A to Town C?

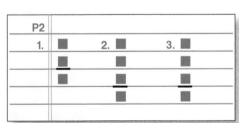

Lesson 56

Part 3

a. $7.88   b. $10.15   c. $3.77   d. $2.89   e. $2.12

1. Don bought two items.
   One item was item A.
   He spent about 10 dollars.
   Which two items did he buy?

2. Ms. Bell bought two items.
   One item was item E.
   She spent about 5 dollars.
   Which two items did she buy?

3. Heather bought two items.
   She spent about 18 dollars.
   Which two items did she buy?

P3					
a.		b.	c.	d.	e.
1.	■ ■		2. ■ ■		3. ■ ■

Independent Work

Part 4
Write 2 times facts and 2 division facts for this number family.

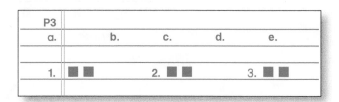

a. $5 \overset{4}{\underset{}{\rightarrow}} 20$

P4		
a.	■ × ■ = ■ ■ × ■ = ■	■ ■ ┐■ ■ ■ ┐■

Lesson

Part 5 For each item, make the number family and work the problem. Write the unit name in the answer.

a. A new tire costs $95.34.
An old tire costs $24.19.
How much more does the new tire cost than the old tire?

b. Fran read 29 fewer pages than Greg read.
Fran read 71 pages.
How many pages did Greg read?

c. Joe had some birdhouses.
Joe sold 72 birdhouses.
He ended up with 121 birdhouses.
How many birdhouses did Joe start with?

d. Joe had 175 more nails than Debbie.
Joe had 340 nails.
How many nails did Debbie have?

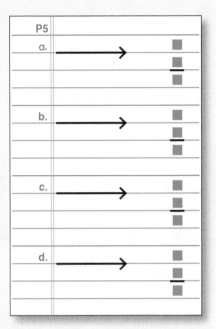

Lesson 56

Part 6 Write each number.

a. six hundred

b. six thousand

c. one thousand ten

d. one thousand one hundred

e. one thousand one

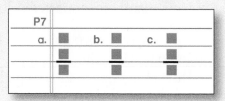

Part 7 For each item, work a column problem to find the missing number.

a. ■ + 21 = 59   b. 47 + ■ = 99   c. ■ + 200 = 750

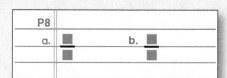

Part 8 Write the fraction for each picture.

a.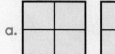

b.

Part 9 Copy and work each problem.

P9			
a.	8000 − 200	b. 4000 −1300	c. 6600 − 990

Part 10 Write the answer to each problem.

a. 34 + 5   b. 92 + 6   c. 44 + 4

d. 83 + 6   e. 25 + 4

P10				
a.	b.	c.	d.	e.

Lesson 57

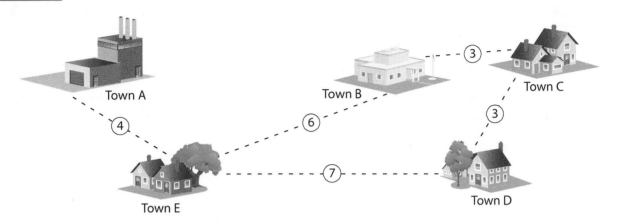

Town A   ④   Town B   ③   Town C   ⑥   ③   ⑦   Town E   Town D

1. There are two routes from Town E to Town C. Figure out how many miles for each route. Circle the answer for the shorter route.

2. There are two routes from Town D to Town B. Figure out how many miles for each route. Circle the answer for the shorter route.

3. There are two routes from Town A to Town C. Figure out how many miles for each route. Circle the answer for the shorter route.

P1			
1. ▪ ▪	2. ▪ ▪	3. ▪ ▪	
▪ ▪	▪ ▪	▪ ▪	
▪ ▪	▪ ▪	▪ ▪	
		▪ ▪	

Part 2

P2				
a. ▪ : ▪	b. ▪ : ▪	c. ▪ : ▪	d. ▪ : ▪	

a.

b.
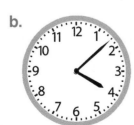

c.

d.

Lesson 57

Part 3

a. $6.42 b. $6.87 c. $2.35 d. $2.79 e. $1.19

1. Chang bought two items. One item was item A.
   He spent about 8 dollars. Which two items did he buy?

2. Ann bought two items. One item was item D.
   She spent about 4 dollars. Which two items did she buy?

3. Tom bought two items. He spent about 3 dollars.
   Which two items did he buy?

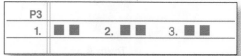

Independent Work

Part 4
For each item, make the number family and work the problem. Write the unit name in the answer.

a. A red shirt cost $11.25.
   A blue shirt cost $14.90
   How much less did the red shirt cost than the blue shirt?

b. Juan bought 95 apples.
   59 of the apples were red apples.
   The rest of the apples were yellow.
   How many yellow apples did Juan buy?

c. There are 245 people on the airplane.
   95 of the people are children.
   How many adults are on the airplane?

d. Joe had 147 dollars less than Tina.
   Joe had 228 dollars.
   How much money did Tina have?

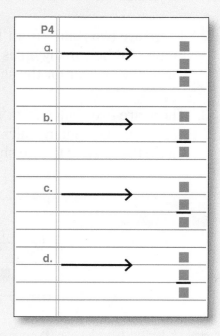

Part 5
Copy each item and write the sign >, <, or =.

P5			
a.	20 + 10 ▮ 5 × 3	b. 5 × 4 ▮ 10 + 10	c. 2 × 4 ▮ 4 × 2

Lesson 57

Part 6

$2.21 $9.26 $1.87 $4.85 $3.91

a. A girl buys the nuts, the hot dog, and the pen. **About** how much does she spend?

b. A girl buys the nuts, the hot dog, and the pen. **Exactly** how much does she spend?

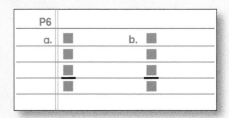

Part 7 | Copy each item and write the sign >, <, or =.

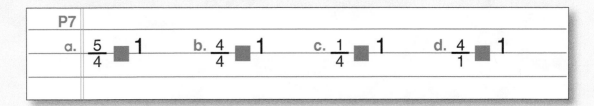

P7

a. $\frac{5}{4}$ ■ 1   b. $\frac{4}{4}$ ■ 1   c. $\frac{1}{4}$ ■ 1   d. $\frac{4}{1}$ ■ 1

Part 8 | Work each problem.

a. How much more is 2450 than 850?

b. How much less is 430 than 900?

Lesson 57

Part 9 Write the answer to each problem.

a. $2\overline{)10}$ b. $2\overline{)8}$ c. $2\overline{)2}$ d. $2\overline{)10}$ e. $2\overline{)4}$

f. $5\overline{)10}$ g. $5\overline{)25}$ h. $2\overline{)10}$ i. $5\overline{)15}$ j. $5\overline{)5}$

P9				
a.	b.	c.	d.	e.
f.	g.	h.	i.	j.

Connecting Math Concepts

Lesson 58

Part 1

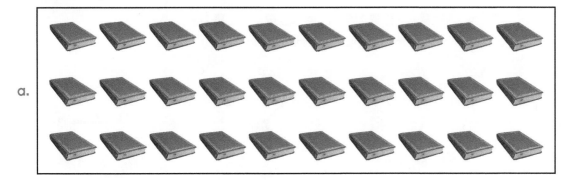

a.

b.

P1	
a.	
b.	

Part 2

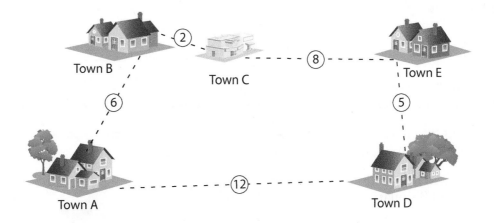

Town B (2) Town C (8) Town E

(6) (5)

Town A (12) Town D

1. There are two routes from Town B to Town D.
    Figure out how many miles for each route.
    Circle the answer for the shorter route.

2. There are two routes from Town A to Town E.
    Figure out how many miles for each route.
    Circle the number for the shorter route.

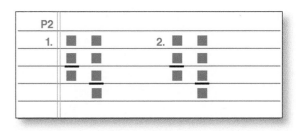

P2		
1.		2.

Lesson 58

Part 3

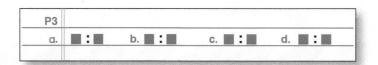

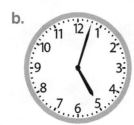  

a.   b.   c.   d.

Part 4

a. $17 - \blacksquare = 3$ b. $45 - \blacksquare = 22$ c. $20 - \blacksquare = 12$

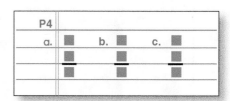

Independent Work

Part 5

a. $1.92   b. $3.91   c. $3.72   d. $2.89   e. $1.42

1. Alice bought two items. One item was item B.
   She spent about 6 dollars. Which two items did she buy?

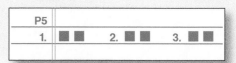

2. Mary bought two items. One item was item A.
   She spent about 5 dollars. Which two items did she buy?

3. Bob bought two items. One item was item E.
   He spent about 3 dollars. Which two items did he buy?

Lesson 58

Part 6
For each item, make the number family and work the problem. Write the unit name in the answer.

a. A man had $18.45 in his pocket.
His pocket tore, and $7.33 fell out of his pocket.
How much did he have left in his pocket?

b. A cow ate 37 more carrots than a goat ate.
The cow ate 148 carrots.
How many carrots did the goat eat?

c. Rita weighed 137 pounds.
Rita weighed 47 pounds less than Henry.
How much did Henry weigh?

d. Rita had 51 pennies.
Rita had 42 fewer pennies than Sarah had.
How many pennies did Sarah have?

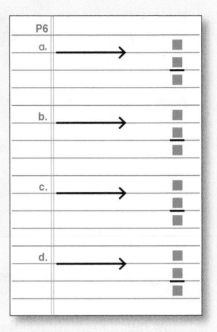

Part 7
For each item, work a column problem to find the missing number.

a. $300 + \blacksquare = 950$

b. $\blacksquare + 20 = 75$

Part 8
Copy and work each regular problem.
Next to the regular problem, work an estimation problem.

Lesson 58

Part 9 Write the dollars and cents amount for each row.

a.

b.

P9			
a.		b.	c.

c.

Part 10 Work each problem.

a. How much more is 2300 than 900?

b. How much less is 1500 than 4000?

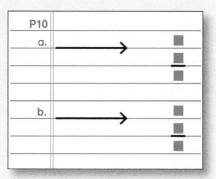

Part 11 Write the answer to each problem.

a. $5\overline{\smash{)}10}$ b. $2\overline{\smash{)}6}$ c. $5\overline{\smash{)}25}$ d. $5\overline{\smash{)}15}$ e. $2\overline{\smash{)}8}$

f. $2\overline{\smash{)}4}$ g. $5\overline{\smash{)}20}$ h. $2\overline{\smash{)}10}$ i. $5\overline{\smash{)}5}$ j. $2\overline{\smash{)}2}$

P11				
a.	b.	c.	d.	e.
f.	g.	h.	i.	j.

Lesson

a. 488   b. 343   c. 132

d. 152   e. 163

P1					
a.	b.	c.	d.	e.	

Part 2

a. 64 – ■ = 23   b. 28 – ■ = 9   c. 74 – ■ = 24

P2		
a. ■	b. ■	c. ■
■	■	■

Part 3

a. A bus started out with 23 people.
Some more people got on the bus.
The bus ended up with 56 people on it.
How many more people got on the bus?

b. A train started out with 70 people.
Then some people got off the train.
The train ended up with 52 people.
How many people got off the train?

c. Jerry had 24 pencils.
He gave away some of the pencils.
He ended up with 14 pencils.
How many pencils did he give away?

Lesson 59

Part 4

a.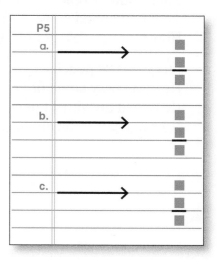

b.

c.

P4	
a.	
b.	
c.	

Part 5

a. There were 125 apples.
   85 of the apples were red.
   The rest of the apples were not red.
   How many apples were not red?

b. There are 125 dogs that are brown
   and 95 dogs that are not brown.
   How many dogs are there in all?

c. The store had 145 shirts.
   65 of the shirts were made of cotton.
   How many shirts were not made of cotton?

P5		
a.	→	▪ ▬ ▪
b.	→	▪ ▬ ▪
c.	→	▪ ▬ ▪

Lesson 59

Independent Work

Part 6 Write the time for each clock.

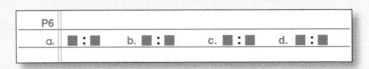

a.  b. 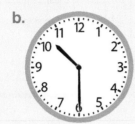 c.  d.

Part 7 Copy each item and write the sign >, <, or =.

P7	
a.	10×3 ■ $20 + 10$ b. $10 - 3$ ■ 4×2 c. $4 + 4$ ■ 3×2

Part 8 For each item, make the number family and work the problem. Write the unit name in the answer.

a. A basketball player is 245 centimeters tall.
A football player is 192 centimeters.
How many centimeters taller is the basketball player than the football player?

b. A redwood tree was 120 feet tall.
The redwood tree was 78 feet taller than the maple tree.
How tall was the maple tree?

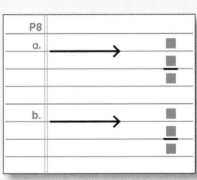

Lesson 59

Part 9
Write **Cu** for cube, **Py** for pyramid, **Sp** for sphere, **H** for hexagon, **P** for pentagon, **R** for rectangle, **S** for square, **T** for triangle.

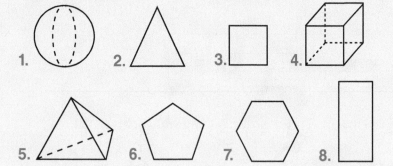

1. 2. 3. 4.

5. 6. 7. 8.

P9			
1.	2.	3.	4.
5.	6.	7.	8.

Part 10

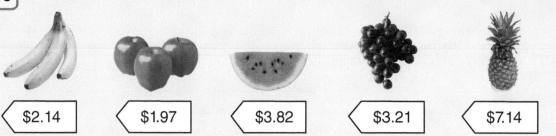

| $2.14 | $1.97 | $3.82 | $3.21 | $7.14 |

a. A girl bought bananas, a watermelon, and a pineapple. **Exactly** how much does she spend?

b. A girl bought bananas, a watermelon, and a pineapple. **About** how much does she spend?

P10			
a.	▪	b.	▪
	▪		▪
	▪		▪
	▪		▪

Lesson

Part 11 Write the dollars and cents amount for each row.

a.

b.

P11		
a.		b.

Part 12 Write the answer to each problem.

a. 5	b. 3	c. 5	d. 2	e. 4
× 5	× 5	× 5	× 2	× 5

P12				
a.	b.	c.	d.	e.
f.	g.	h.	i.	j.

f. 2	g. 2	h. 1	i. 3	j. 4
× 2	× 5	× 5	× 2	× 2

Lesson 60

Part 1

a.

b.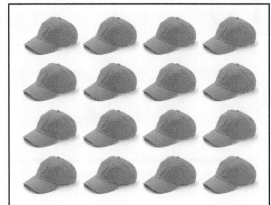

c.

P1	
a.	
b.	
c.	

Part 2

P2	
a. $\dfrac{3}{4} + \dfrac{2}{4} = $ ___	b. $\dfrac{9}{6} - \dfrac{5}{6} = $ ___
c. $\dfrac{5}{3} + \dfrac{4}{3} = $ ___	d. $\dfrac{6}{5} - \dfrac{1}{5} = $ ___

Part 3

a. ■ − 12 = 30   b. ■ − 15 = 40   c. 95 − ■ = 23

P3			
a. ■	b. ■	c. ■	
■	■	■	
■	■	■	

Lesson

Part 4

a. A boy had 24 books.
   Then his sister gave him more books.
   He ended up with 36 books.
   How many books did his sister give him?

b. A boy had some books.
   He gave his sister 14 books.
   He ended up with 35 books.
   How many books did he start with?

c. A teacher had 36 pencils.
   She gave away some pencils.
   She ended up with 21 pencils.
   How many pencils did she give away?

d. The train started with 289 passengers.
   35 more passengers got on the train.
   How many passengers did the train end up with?

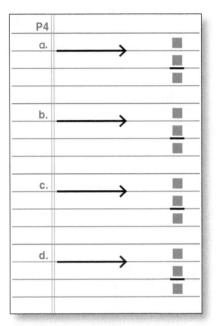

Independent Work

Part 5 Round each number to the nearest hundred.

a. 462   b. 629   c. 318

d. 193   e. 458

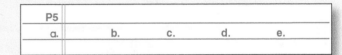

Part 6 Write the time for each clock.

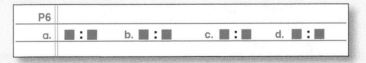

a.

b.

c.
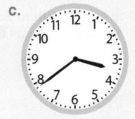

d.
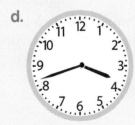

Lesson 60

Part 7 For each item, make the number family and work the problem. Write the unit name in the answer.

a. Jerry had 96 coins.
23 of the coins were dimes.
How many of the coins were not dimes?

b. A cow ate 97 more pounds of hay than a horse ate.
The horse ate 96 pounds of hay.
How many pounds of hay did the cow eat?

c. A lion weighed 564 pounds.
A bear weighed 820 pounds.
How much more did the bear weigh than the lion?

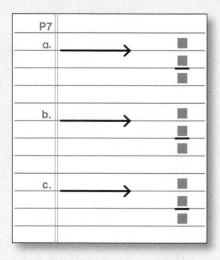

Part 8 Write each number.

a. two thousand

b. two hundred

c. two hundred ten

d. two thousand ten

e. one thousand two hundred thirteen

Part 9 Work each problem.

a. $6.87   b. $1.05   c. $3.10   d. $4.94   e. $5.92

1. Carlos bought two of these items. One item was item B.
He spent about 6 dollars. Which two items did he buy?

2. Bob bought two items. One item was item A.
He spent about 13 dollars. Which two items did he buy?

3. Tom bought two items. One item was item A.
He spent about 12 dollars. Which two items did he buy?

Part 10 For each item, work the column problem to find the missing number.

a. ■ + 240 = 900   b. 1500 + ■ = 3500

P10		
a.	■	b. ■
	■	■
	■	■

Part 11 Copy each item and write the sign >, <, or =.

P11				
a. $\frac{6}{6}$ ■ 1	b. $\frac{6}{7}$ ■ 1	c. $\frac{7}{6}$ ■ 1	d. $\frac{1}{6}$ ■ 1

Lesson 61

Part 1

a. Mike had some friends.
   14 friends moved away.
   Now Mike has 25 friends.
   How many friends did Mike start with?

b. Tina started with 25 seashells.
   She found some more seashells.
   Now Tina has 47 seashells.
   How many seashells did she find?

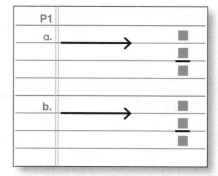

Part 2

a. ■ − 24 = 70   b. 52 − ■ = 39   c. ■ − 24 = 12

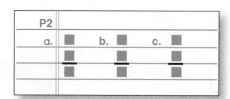

Part 3

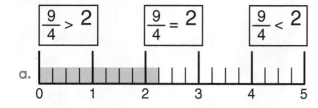

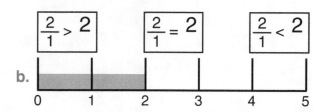

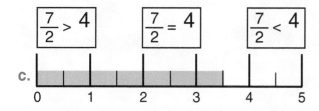

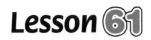

Lesson 61

Independent Work

Part 4 Copy and work each problem.

P4

a. $\frac{5}{7} + \frac{1}{7} = \frac{\quad}{\quad}$ b. $\frac{5}{3} - \frac{2}{3} = \frac{\quad}{\quad}$ c. $\frac{4}{10} + \frac{3}{10} = \frac{\quad}{\quad}$

Part 5 For each item, make the number family and work the problem. Write the unit name in the answer.

a. The boat was 12 feet shorter than the truck.
The boat was 42 feet long.
How many feet long was the truck?

b. Wet shirts and dry shirts were on the line.
There were 23 shirts altogether.
7 of the shirts were wet.
How many shirts were dry?

c. There were 52 children in the park.
31 children were playing.
The rest were eating.
How many children were eating?

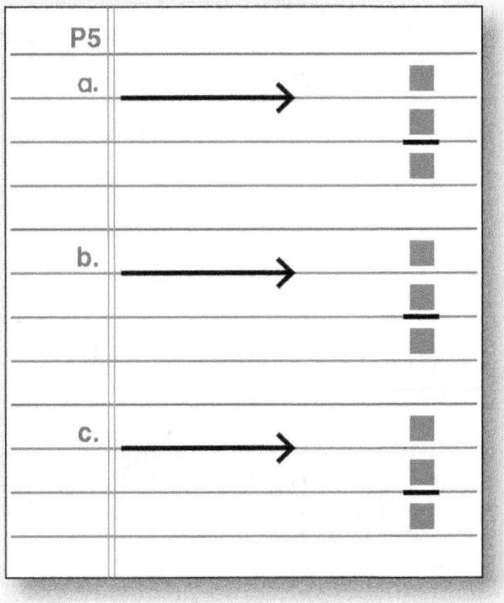

Part 6 For each item, write the column problem to find the missing number.

a. 150 + ■ = 350 b. 150 + 350 = ■ c. ■ + 15 = 35

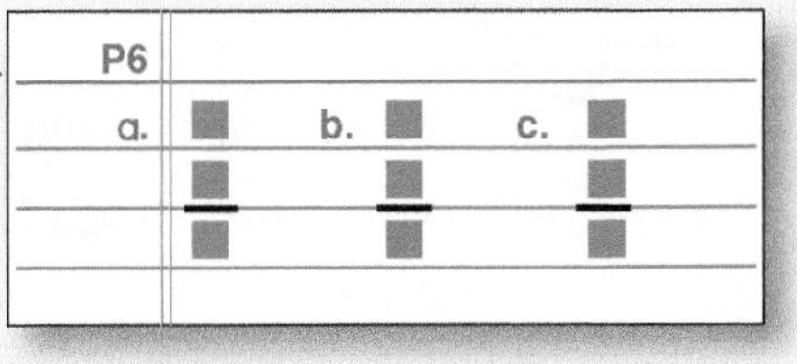

Lesson 61

Part 7

$2.21 $9.26 $1.87 $4.85 $3.95

a. A man buys the hat, the book, and the pen.
Exactly how much does he spend?

b. A man buys the hat, the book, and the pen.
About how much does he spend?

P7		
a.	■	b. ■
	■	■
	■	■
	■	■

Part 8 Round each number to the nearest hundred.

a. 138   b. 172   c. 259
d. 417   e. 261

P8					
a.	b.	c.	d.	e.

Part 9 Write the dollars and cents amount for each row.

a.

b.

P9		
a.	b.	c.

c.

Lesson 61

Part 10 Copy and work each problem.

a.
```
   199
 +  99
```
b.
```
  3550
 + 550
```
c.
```
  1288
 +   88
```

P10			
a. ■	b. ■	c. ■	
▬	▬	▬	
■	■	■	

Part 11 Write the numbers for counting by 4s to 40 and 9s to 90.

P11		
a.	4 ■ ■ ■ ■ ■ ■ ■	40
b.	9 ■ ■ ■ ■ ■ ■ ■	90

Part 12 Write the answer to each problem.

a. 4⟌20   b. 4⟌12   c. 4⟌4   d. 4⟌16   e. 4⟌8

f. 2⟌18   g. 4⟌16   h. 5⟌20   i. 4⟌12   j. 5⟌5

P12				
a.	b.	c.	d.	e.
f.	g.	h.	i.	j.

Lesson 62

Part 1

AM   PM

Midnight   Noon   Midnight
12 1 2 3 4 5 6 7 8 9 10 11 12 1 2 3 4 5 6 7 8 9 10 11 12

a. Kara drove to Hill Town.
She left at 6 AM and drove 7 hours.
When did she arrive in Hill Town?

b. Oscar left on a trip at 5 AM to Chicago.
He drove for 6 hours.
When did he arrive in Chicago?

P1		
a.	b.	c.

c. Sarah left for Chicago at 3 AM.
She drove for 12 hours.
When did she arrive in Chicago?

Part 2

a. $\blacksquare - 13 = 80$ b. $48 - \blacksquare = 23$ c. $\blacksquare - 80 = 7$

d. $\blacksquare - 28 = 24$ e. $34 - \blacksquare = 19$

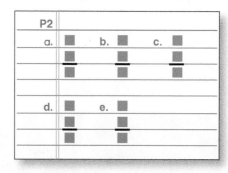

Part 3

a. $3000 + 200 + 10 + 5$ b. $1000 + 400 + 60 + 2$

c. $7000 + 300 + 40 + 1$ d. $1000 + 200 + 10 + 4$

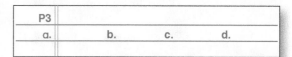

Lesson 62

Part 4

a. The bus started out with 11 people on it.
   Some more people got on the bus.
   Now the bus has 42 people on it.
   How many people got on the bus?

b. There were some books on the shelf yesterday.
   Tim took 13 books off the shelf today.
   Now there are 48 books on the shelf.
   How many books were on the shelf yesterday?

c. There were 16 books on the shelf yesterday.
   Today there are 4 books on the shelf.
   How many books were taken from the shelf?

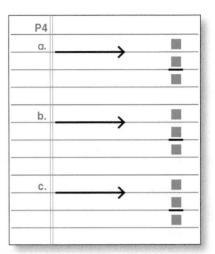

Independent Work

Part 5 | For each item, write the statement that tells about the number line.

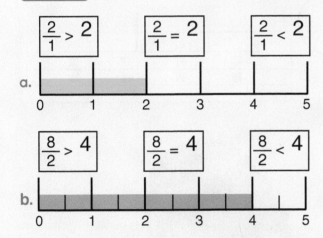

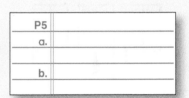

Part 6 | Round each number to the nearest hundred.

a. 182   b. 238   c. 529

d. 274   e. 762

P6					
a.		b.	c.	d.	e.

Part 7 | Copy and work each problem.

a.   4 5 6
     + 1 7 9

b.   3 9 9 0
     + 1 5 5 0

c.   8 8 8
     + 7 7 7

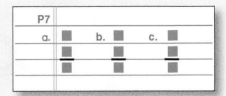

Part 8 For each item, make the number family and work the problem. Write the unit name in the answer.

a. The red squirrel ate 23 more peanuts than the gray squirrel ate.
The red squirrel ate 31 peanuts.
How many peanuts did the gray squirrel eat?

b. Jackie ran 56 miles during the week.
Debbie ran 40 miles during the week.
How much farther did Jackie run than Debbie ran?

c. There were long worms and short worms in the can.
42 worms were short. 24 worms were long.
How many worms were in the can?

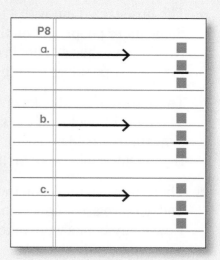

Part 9 Write the fraction for each picture.

a.

b.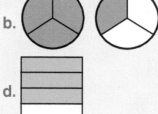

c.

d.

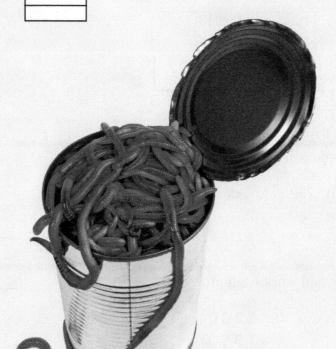

Lesson 62

Part 10

a. $6.85 b. $6.05 c. $1.62 d. $1.12 e. $2.64

1. Jerry bought two of these items. One item was item B.
   He spent about 7 dollars. Which two items did he buy?

2. Bob bought two items. One item was item A.
   He spent about 8 dollars. Which two items did he buy?

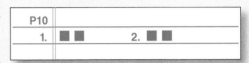

Part 11 | For each item, write the column problem to find the missing number.

a. $30 + 90 = \blacksquare$ b. $\blacksquare + 20 = 70$ c. $100 + \blacksquare = 175$

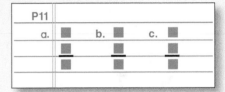

Part 12 | Write each number.

a. nine thousand

b. nine hundred

c. nine thousand sixty

d. two thousand forty-four

Lesson 63

Part 1

AM   PM

| Midnight | | | | | | | | | | | Noon | | | | | | | | | | | Midnight |
| 12 | 1 | 2 | 3 | 4 | 5 | 6 | 7 | 8 | 9 | 10 | 11 | 12 | 1 | 2 | 3 | 4 | 5 | 6 | 7 | 8 | 9 | 10 | 11 | 12 |

a. Tina started at 7 AM. She flew to New York.
The trip took 8 hours. What time did she arrive in New York?

b. Alberto flew from Denver to Houston. He left Denver at 11 AM.
The trip took 3 hours. What time did he arrive in Houston?

c. Rose flew from California to Mexico. The trip took 5 hours.
The plane left California at 6 PM. What time did she arrive
in Mexico?

d. Al walked from Lakeland to Small Town.
He started at 7 AM and walked 11 hours.
What time did he arrive in Small Town?

P1				
a.		b.	c.	d.

Part 2

a. $30 - \blacksquare = 12$ b. $\blacksquare - 27 = 21$ c. $\blacksquare - 14 = 32$

d. $42 - \blacksquare = 25$ e. $\blacksquare - 52 = 34$

P2			
a. ■	b. ■	c. ■	
■	■	■	
▬	▬	▬	
d. ■	e. ■		
■	■		
▬	▬		

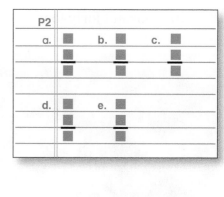

   Connecting Math Concepts

Lesson 63

Part 3

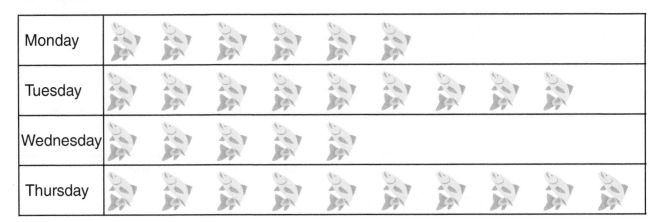

Monday	🐟🐟🐟🐟🐟🐟
Tuesday	🐟🐟🐟🐟🐟🐟🐟🐟🐟
Wednesday	🐟🐟🐟🐟🐟
Thursday	🐟🐟🐟🐟🐟🐟🐟🐟🐟🐟

a. How many fish were caught on Monday?

b. On how many days were more than 5 fish caught?

c. On which day were the most fish caught?

d. How many fish were caught on that day?

e. On which day were the fewest fish caught?

f. How many fish were caught on that day?

P3		
a.	b.	c.
d.	e.	f.

Part 4

a. The train started out with 138 passengers.
Some passengers got off the train.
The train ended up with 49 passengers.
How many passengers got off the train?

b. Donald started out with some money.
He spent $2.36.
He ended up with $8.14.
How much money did he start with?

c. Mr. Broom had some cows.
He bought 48 more cows.
Now he has 132 cows.
How many cows did he start with?

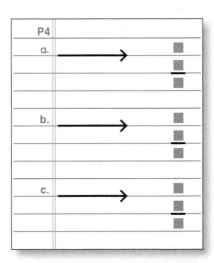

Lesson 63

Part 5 Write the answer to each place-value problem.

a. 4000 + 800 + 10 + 2   b. 2000 + 400 + 70 + 5

c. 1000 + 400 + 60 + 7   d. 9000 + 400 + 10 + 3

P5				
a.		b.	c.	d.

Part 6 For each item, write the statement that tells about the number line.

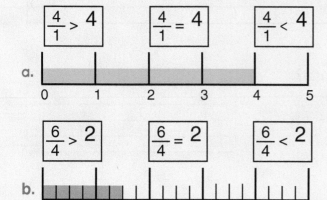

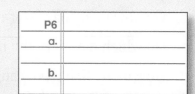

P6	
a.	
b.	

Part 7 Write the time for each clock.

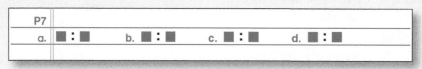

P7				
a. ■:■	b. ■:■	c. ■:■	d. ■:■	

a.    b.    c.    d.

Part 8 Copy and work each problem.

a. 5200
   −1750

b. 400
   − 60

c. 5000
   − 500

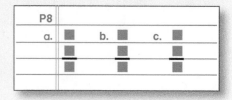

P8			
a. ■	b. ■	c. ■	
■	■	■	
■	■	■	

Lesson 63

Part 9

a. Find the perimeter of this rectangle.

b. Find the area of this rectangle.

9 ft

3 ft

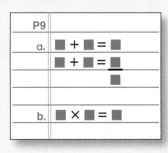

P9	
a.	■ + ■ = ■
	■ + ■ = ■
	■
b.	■ × ■ = ■

Part 10 Round each number to the nearest hundred.

a. 539   b. 781   c. 219
d. 653   e. 471

P10					
a.		b.	c.	d.	e.

Part 11 Write **Cu** for cube, **Py** for pyramid, **Sp** for sphere, **H** for hexagon, **P** for pentagon, **R** for rectangle, **S** for square, **T** for triangle.

1.

2.

3.

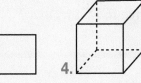
4.

P11				
1.	2.	3.	4.
5.	6.	7.	8.

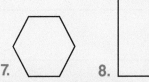
5.   6.   7.   8.

Lesson 64

Part 1

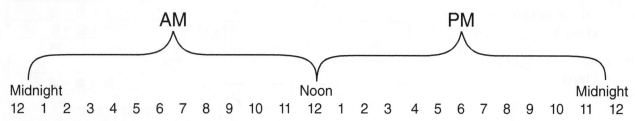

a. Marcos started at 7 AM. He drove to New York.
The trip took 8 hours. When did he arrive in New York?

b. Jerry started working at 8 AM. He worked 8 hours.
What time did he finish working?

c. The man started walking at 10 AM. He walked for
5 hours.
What time did he stop walking?

d. Tom drove from Small Town to River Town.
He left Small Town at 9 AM.
He drove for 7 hours. When did he arrive in River Town?

P1				
a.		b.	c.	d.

Part 2

January	☁ ☁ ☁
February	☁ ☁ ☁ ☁ ☁ ☁ ☁
March	☁
April	☁ ☁ ☁ ☁ ☁
May	

a. Which month had no cloudy days?

b. Which months had more than 3 cloudy days?

c. Which month had the most cloudy days?

d. Which month had the fewest cloudy days?

P2	
a.	b.
c.	d.

Lesson  64

Part 3

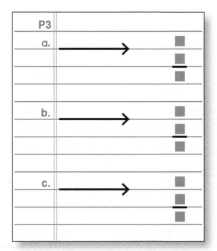

a. Dan began the day with 74 seashells.
    He found some more seashells during the day.
    He ended up with 92 seashells.
    How many seashells did Dan find?

b. Shen started out with some money.
    He bought a pen that cost $2.08.
    Now Shen has $7.92.
    How much money did Shen start with?

c. The class began the day with 95 points.
    The class earned lots of points during the day.
    At the end of the day, the class had 146 points.
    How many points did the class earn?

Independent Work

Part 4 For each item, work a column problem to find the missing number.

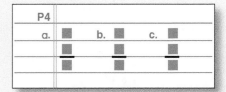

a. $80 - \blacksquare = 50$ b. $\blacksquare - 20 = 10$ c. $70 - 60 = \blacksquare$

Part 5 Write the answer to each place-value problem.

a. $3000 + 600 + 10 + 4$ b. $2000 + 900 + 40 + 2$

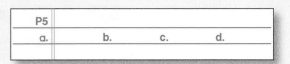

c. $7000 + 100 + 20 + 8$ d. $1000 + 100 + 10 + 1$

Part 6 Copy and work each problem.

a.
$$\begin{array}{r} 311 \\ -199 \\ \hline \end{array}$$
b.
$$\begin{array}{r} 715 \\ -39 \\ \hline \end{array}$$
c.
$$\begin{array}{r} 522 \\ -88 \\ \hline \end{array}$$

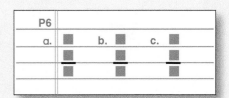

Lesson 64

Part 7

| $1.09 | $2.93 | $6.11 | $2.06 | $3.88 |

a. A boy buys cereal and juice.
   About how much does he spend?

b. A boy buys cereal and juice.
   Exactly how much does he spend?

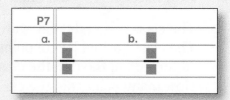

Part 8 Round each number to the nearest hundred.

a. 137   b. 186   c. 374
d. 482   e. 718

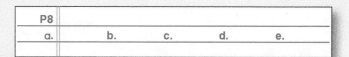

Part 9 Write the time for each clock.

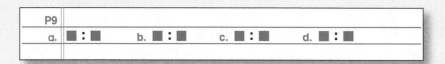

a.   b.   c.   d.

Connecting Math Concepts

Lesson 65

Part 1

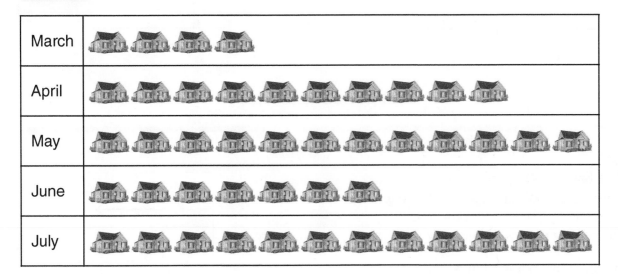

March	
April	
May	
June	
July	

a. Were more houses built in March or April?

b. In 2 months, the same number of houses were built. What are those 2 months?

c. In which month were the fewest houses built?

d. In how many months were more than 3 houses built?

P1	
a.	b.
c.	d.

Part 2

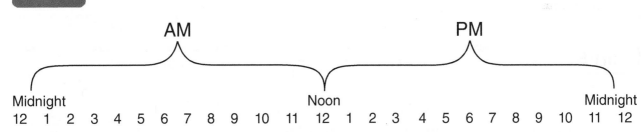

AM   PM

| Midnight | | | | | | | | | | | Noon | | | | | | | | | | | Midnight |
| 12 | 1 | 2 | 3 | 4 | 5 | 6 | 7 | 8 | 9 | 10 | 11 | 12 | 1 | 2 | 3 | 4 | 5 | 6 | 7 | 8 | 9 | 10 | 11 | 12 |

a. Jane started at 9 AM. She flew to Boston. The trip took 6 hours. What time did she arrive in Boston?

b. Dan flew from Dallas to Chicago. He left Dallas at 10 AM. The trip took 3 hours. What time did he arrive in Chicago?

c. Al flew from Mexico to Texas. The trip took 5 hours. The plane left Mexico at 6 PM. What time did it arrive in Texas?

d. Melissa walked from Springfield to Scottsville. She started at 9 AM and walked for 7 hours. What time did she arrive in Scottsville?

P2			
a.	b.	c.	d.

Connecting Math Concepts

Lesson

a. $\begin{array}{r} 649 \\ +210 \\ \hline \end{array}$ b. $\begin{array}{r} 565 \\ +237 \\ \hline \end{array}$ c. $\begin{array}{r} 487 \\ +392 \\ \hline \end{array}$

P3			
a.		b.	c.

Independent Work

Part 4 | Work each problem. Write the unit name.

a. The bag had 45 grapes in it.
   Somebody took grapes from the bag.
   Now the bag has 13 grapes in it.
   How many grapes did somebody take?

b. Amanda had some money.
   She spent 37 dollars.
   Now she has 63 dollars.
   How many dollars did Amanda start with?

c. The truck weighed 4826 pounds.
   The car weighed 2942 pounds.
   How many pounds more did the truck weigh than the car?

d. Tom weighed 145 pounds.
   Tom weighed 25 pounds less than Jerry.
   How many pounds did Jerry weigh?

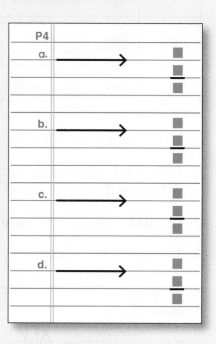

Part 5 | For each item, work the column problem to find the missing number.

a. 90 − ■ = 30   b. ■ − 20 = 58   c. 30 + ■ = 80

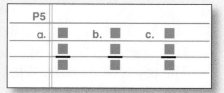

Part 6 | Write the numbers for counting by 4s to 40 and 3s to 30.

P6		
a.	4 ■ ■ ■ ■ ■ ■ ■	40
b.	3 ■ ■ ■ ■ ■ ■ ■	30

Lesson

Part 7 Copy each item and write the sign >, <, or =.

P7	
a. 4500 ■ 3500	b. 4100 ■ 4001	c. 2035 ■ 2350

Part 8 Write the answer to each place-value problem.

a. 1000 + 200 + 10 + 2   b. 6000 + 100 + 30 + 1

c. 2000 + 400 + 40 + 1   d. 1000 + 900 + 10 + 4

P8			
a.	b.	c.	d.

Part 9 Write the fraction for each picture.

a.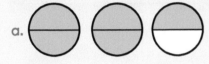

b.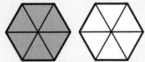

c.

d.

P9			
a. ▪/▪	b. ▪/▪	c. ▪/▪	d. ▪/▪

Part 10 Copy each problem and write the answer.

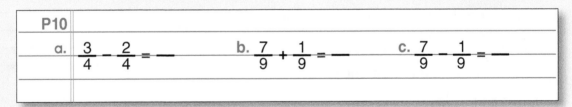

P10			
a. $\dfrac{3}{4} - \dfrac{2}{4} = \overline{}$	b. $\dfrac{7}{9} + \dfrac{1}{9} = \overline{}$	c. $\dfrac{7}{9} - \dfrac{1}{9} = \overline{}$

Part 11

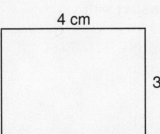

4 cm

3 cm

a. Find the perimeter of this rectangle.

b. Find the area of this rectangle.

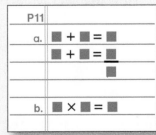

P11	
a.	■ + ■ = ■
	■ + ■ = $\dfrac{■}{■}$
b.	■ × ■ = ■

Lesson 66

Part 1

a. $\begin{array}{r} 563 \\ +298 \\ \hline \end{array}$ b. $\begin{array}{r} 234 \\ +454 \\ \hline \end{array}$ c. $\begin{array}{r} 489 \\ -223 \\ \hline \end{array}$ d. $\begin{array}{r} 345 \\ +367 \\ \hline \end{array}$

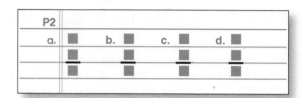

Part 2

a. How many hours is it from 5 PM to midnight?

b. How many hours is it from 2 AM to noon?

c. How many hours is it from 6 PM to midnight?

d. How many hours is it from 3 PM to midnight?

Independent Work

Part 3 For each item, make the number family and work the problem. Write the unit name in the answer.

a. The truck weighs 473 pounds more than the car.
The car weighs 1931 pounds.
How much does the truck weigh?

b. There were 126 fewer people in the park
than there were at the beach.
781 people were at the beach.
How many people were in the park?

c. Jane had lots of crayons.
She gave away 42 crayons.
She ended up with 62 crayons.
How many crayons did she start with?

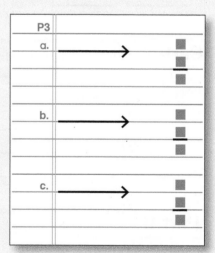

Connecting Math Concepts

Part 4 Copy each item and write the sign >, <, or =.

P4			
a. $\dfrac{7}{7}$ ■ $\dfrac{3}{2}$	b. $\dfrac{4}{5}$ ■ $\dfrac{3}{3}$	c. $\dfrac{4}{3}$ ■ $\dfrac{2}{2}$	

Part 5 For each item, write the time the person arrived.
Remember to write AM or PM in your answer.

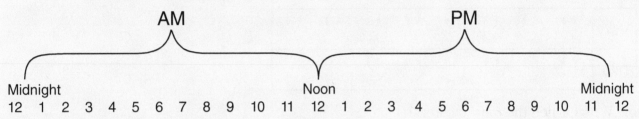

AM   PM

Midnight   Noon   Midnight
12 1 2 3 4 5 6 7 8 9 10 11 12 1 2 3 4 5 6 7 8 9 10 11 12

a. Jerry started a trip at 9 AM. His trip took 4 hours.
At what time did he arrive?

b. Maria started a trip at 10 AM. Her trip took 5 hours.
At what time did she arrive?

P5		
a.	b.	c.

c. Joe started a trip at 7 PM. His trip took 4 hours.
At what time did he arrive?

Part 6 Write the answer to each question.

This graph shows the number of happy face stickers different students earned.

Jerry	😊 😊 😊 😊
Elisa	😊 😊 😊 😊 😊 😊 😊
Tom	😊 😊 😊
Mary	😊 😊 😊 😊 😊 😊 😊 😊
Tina	😊 😊 😊 😊 😊 😊

😊 = 1 sticker

P6		
a.		b.
c.		d.

a. Who earned the most stickers?

b. How many students earned more than 4 stickers?

c. Who earned the fewest stickers?

d. Who earned fewer than 4 stickers?

Part 7 Write the answer to each place-value problem.

a. 5000 + 500 + 10 + 5   b. 3000 + 900 + 10 + 2

c. 4000 + 600 + 20 + 1   d. 1000 + 100 + 90 + 9

P7			
a.	b.	c.	d.

Part 8 For each item, work the column problem to find the missing number.

a. ■ − 30 = 70   b. 90 − ■ = 30   c. ■ − 30 = 50

P8		
a.	b.	c.

Part 9 Copy and work each problem.

P9	
a. $\dfrac{7}{9} + \dfrac{1}{9} = $ ___   b. $\dfrac{5}{7} - \dfrac{2}{7} = $ ___   c. $\dfrac{3}{9} + \dfrac{5}{9} = $ ___

Connecting Math Concepts

Lesson 66

Part 10 Write the answer to each problem.

a. 4
  ×4

b. 3
  ×3

c. 4
  ×3

d. 2
  ×4

e. 3
  ×4

f. 3
  ×5

g. 3
  ×2

h. 4
  ×4

i. 3
  ×3

j. 4
  ×5

P10					
a.		b.	c.	d.	e.
f.		g.	h.	i.	j.

Lesson 67

Part 1

a. 4251   b. 1892   c. 3412

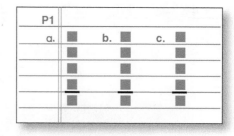

Part 2

a. How many hours is it from 2 PM to midnight?

b. How many hours is it from 8 AM to noon?

c. How many hours is it from 5 PM to midnight?

d. How many hours is it from 4 AM to noon?

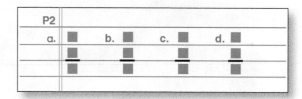

Part 3

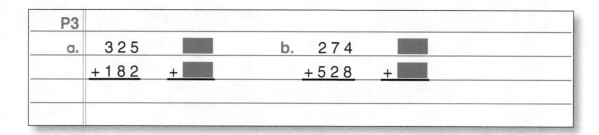

P3			
a.	325 +182	▮ +▮	b. 274 +528

Independent Work

Part 4 For each item, make the number family and work the problem. Write the unit name in the answer.

a. Sal weighs 237 pounds.
   Jim weighs 157 pounds.
   How many pounds heavier is Sal than Jim?

b. A redwood tree was 120 feet tall.
   The redwood tree was 78 feet taller than a maple tree.
   How tall was the maple tree?

c. Doris had lots of seashells. She gave away 42 of the seashells. She ended up with 58 seashells.
   How many seashells did she start with?

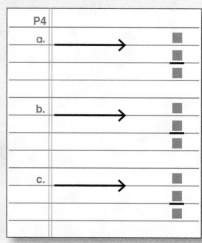

Part 5 | Copy each item and write the sign >, <, or =.

P5				
a.	$\dfrac{3}{3}$ ▪ $\dfrac{9}{10}$	b. $\dfrac{2}{7}$ ▪ $\dfrac{4}{4}$	c. $\dfrac{5}{5}$ ▪ $\dfrac{3}{2}$	d. $\dfrac{8}{8}$ ▪ $\dfrac{5}{5}$

Part 6 | Write the answer to each place-value problem.

a. 2000 + 100 + 10 + 9   b. 1000 + 400 + 40 + 1

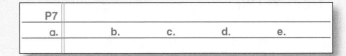

P6			
a.	b.	c.	d.

c. 7000 + 700 + 70 + 7   d. 4000 + 300 + 10 + 1

Part 7 | Round each number to the nearest hundred.

a. 194   b. 273   c. 319

P7				
a.	b.	c.	d.	e.

d. 238   e. 426

Part 8 | Write the fraction for each picture.

a.

b.

c.

d.

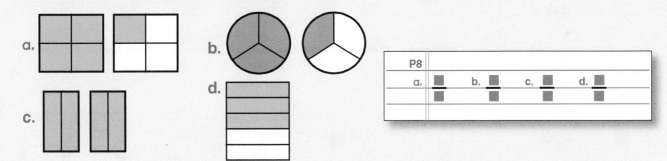

P8			
a. ▪/▪	b. ▪/▪	c. ▪/▪	d. ▪/▪

Part 9 | For each item, write the column problem to find the missing number.

a. ▪ + 20 = 100   b. 200 + ▪ = 250

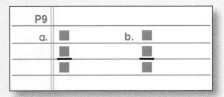

P9	
a. ▪/▪/▪	b. ▪/▪/▪

Lesson 67

Part 10 For each item, write the time the person arrived.
Remember to write AM or PM in your answer.

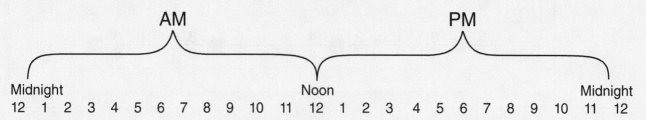

AM   PM

Midnight   Noon   Midnight
12 1 2 3 4 5 6 7 8 9 10 11 12 1 2 3 4 5 6 7 8 9 10 11 12

a. Tom started a trip at 6 PM. His trip took 4 hours.
When did he arrive?

b. Mr. James started a trip at 6 AM. His trip took 5 hours.
At what time did he arrive?

P10			
a.		b.	c.

c. Melissa started a trip at 9 AM. Her trip took 6 hours.
At what time did she arrive?

Part 11 Write the answer to each problem.

a. 4   b. 3   c. 0   d. 4   e. 0
   ×0   ×4   ×3   ×4   ×4

f. 3   g. 0   h. 3   i. 6   j. 6
   ×3   ×2   ×1   ×0   ×1

P11					
a.	b.	c.	d.	e.
f.	g.	h.	i.	j.

Part 12 Work the problems to find the missing numbers.

a. ■ − 40 = 20   b. 40 − ■ = 10   c. ■ − 10 = 50

P12			
a. ■	b. ■	c. ■	
■	■	■	
■	■	■	

Lesson  68

Part 1

a. How many hours is it from 7 AM to noon?

b. How many hours is it from 10 PM to midnight?

c. How many hours is it from 4 PM to midnight?

d. How many hours is it from 6 AM to noon?

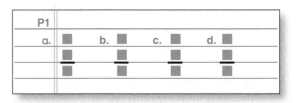

Independent Work

Part 2 Write the place-value addition fact for each number.

a. 3241   b. 9814   c. 1819

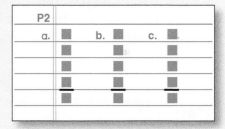

Part 3 Copy and work each regular problem. Next to the regular problem, work an estimation problem with the numbers rounded to the nearest hundred.

P3				
a.	4 2 7	▇	b. 5 8 6	▇
	+3 7 2	+ ▇	+2 1 3	+ ▇

Lesson 68

Part 4 For each item, make the number family and work the problem. Write the unit name in the answer.

a. The store had birdhouses.
The store sold 72 birdhouses.
The stored ended up with 121 birdhouses.
How many birdhouses did the store start with?

b. Joe had 420 nails.
He used 107 nails to build a chair.
How many nails did he have left?

c. Elva was 14 centimeters shorter than Rose.
Rose was 148 centimeters tall. How tall was Elva?

d. Jason went walking on Wednesday and Thursday.
He walked a total of 136 minutes.
He walked for 95 minutes on Wednesday.
How many minutes did he walk on Thursday?

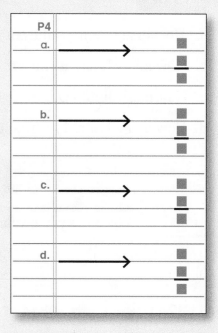

Part 5 Write the answer to each problem.

a. $\begin{array}{r} 0 \\ \times 4 \\ \hline \end{array}$ b. $\begin{array}{r} 4 \\ \times 4 \\ \hline \end{array}$ c. $\begin{array}{r} 1 \\ \times 4 \\ \hline \end{array}$ d. $\begin{array}{r} 3 \\ \times 4 \\ \hline \end{array}$ e. $\begin{array}{r} 3 \\ \times 3 \\ \hline \end{array}$

f. $\begin{array}{r} 0 \\ \times 3 \\ \hline \end{array}$ g. $\begin{array}{r} 3 \\ \times 2 \\ \hline \end{array}$ h. $\begin{array}{r} 3 \\ \times 5 \\ \hline \end{array}$ i. $\begin{array}{r} 4 \\ \times 3 \\ \hline \end{array}$ j. $\begin{array}{r} 4 \\ \times 2 \\ \hline \end{array}$

P5				
a.	b.	c.	d.	e.
f.	g.	h.	i.	j.

Part 6 Write the time for each clock.

P6				
a. ■ : ■	b. ■ : ■	c. ■ : ■	d. ■ : ■

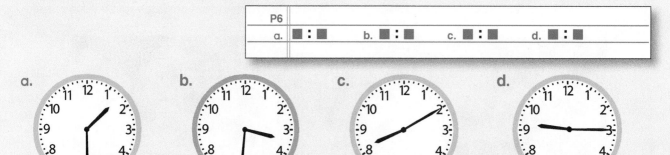

Part 7 Copy and work each problem.

a. $\begin{array}{r} 320 \\ -\ 85 \end{array}$

b. $\begin{array}{r} 4151 \\ -\ 909 \end{array}$

c. $\begin{array}{r} 3000 \\ -\ 200 \end{array}$

P7			
a. ■	b. ■	c. ■	
▬	▬	▬	
■	■	■	

Part 8

a. Find the perimeter of this rectangle.

b. Find the area of this rectangle.

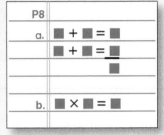

5 cm

2 cm

P8	
a. ■ + ■ = ■	
■ + ■ = ▬	
■	
b. ■ × ■ = ■	

Part 9

$2.14 $1.87 $3.82 $3.11 $7.14

a. A girl bought bananas, apples, and grapes. **Exactly** how much did she spend?

b. A girl bought bananas, apples, and grapes. **About** how much did she spend?

P9		
a. ■	b. ■	
■	■	
▬	▬	

Lesson 69

Independent Work

Part 1 | Work each problem. Write the unit name.

a. The horse is 125 pounds lighter than the cow.
The horse weighs 382 pounds.
How many pounds does the cow weigh?

b. A man had some money.
Then he spent $6.50.
He ended up with $5.19.
How much money did he start out with?

c. The car went 1420 miles.
The truck went 2305 miles.
How many more miles did the truck go than the car?

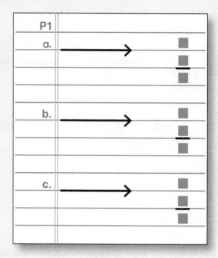

Part 2 | Write the place-value addition fact for each number.

a. 1625　　b. 2341

Part 3 | Copy each item and write the sign >, <, or =.

P3			
a. $\dfrac{6}{6}$ ▉ $\dfrac{9}{4}$	b. $\dfrac{6}{5}$ ▉ $\dfrac{9}{9}$	c. $\dfrac{4}{4}$ ▉ $\dfrac{8}{8}$

Part 4 | Copy and work each regular problem. Next to the regular problem, work an estimation problem with the numbers rounded to the nearest hundred.

P4

a.
```
  3 1 7        ▉
+ 4 0 3     + ▉
```

b.
```
  2 9 1        ▉
+ 4 8 6     + ▉
```

　　　　　　　　　　　　　　　　　　Connecting Math Concepts

Lesson 69

Part 5 Write the answer to each problem.

a. 5
×8

b. 9
×5

c. 6
×5

d. 5
×7

e. 0
×5

f. 4
×3

g. 3
×3

h. 4
×4

i. 3
×4

j. 5
×9

P5				
a.	b.	c.	d.	e.
f.	g.	h.	i.	j.

Part 6 Copy and work each problem.

a. 222
− 77

b. 324
− 89

c. 5050
− 909

P6			
a. ■	b. ■	c. ■	
■	■	■	
■	■	■	

Part 7 Copy each item and write the sign >, <, or =.

P7		
a.	2060 ■ 2600	b. 1900 ■ 1090

Part 8 Write the fraction for each number line.

a.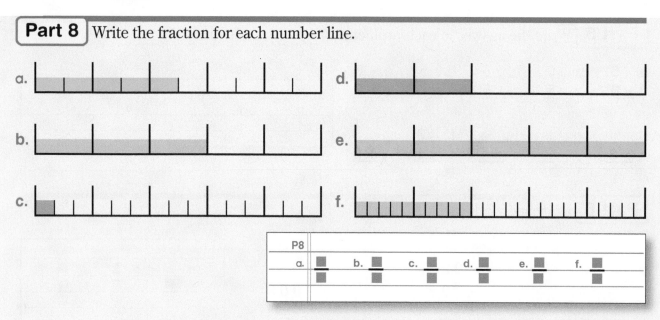

b.

c.

d.

e.

f.

P8						
a. ■	b. ■	c. ■	d. ▬	e. ■	f. ▬	
■	■	■	■	■	■	

Part 9 Copy each item and write the sign >, <, or =.

P9	
a. $10 - 1$ ■ 5×2	b. 3×5 ■ $7 + 7$

Lesson

a. Each dime is worth 10 cents.

b. There are 15 horses in each barn.

c. Each car has 3 people in it.

d. There are 24 hours in each day.

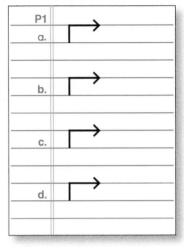

Part 2

a. 8 ⌐ P → 96   b. 9 ⌐ 81 → Y   c. K ⌐ 9 → 90

d. 5 ⌐ J → 50   e. 8 ⌐ 80 → H   f. M ⌐ 5 → 40

P2		
a.	b.	c.
d.	e.	f.

Part 3

a. How many hours is it from 11 AM to 9 PM?

b. How many hours is it from 3 AM to 4 PM?

c. How many hours is it from 10 PM to 1 AM?

Part 4

P4		
a. 3 2 × 4	b. 1 3 2 × 3	c. 3 1 0 × 2

Lesson 70

Independent Work

Part 5 Write the place-value addition fact for each number.

a. 3729   b. 4236

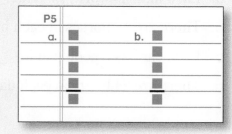

Part 6 Copy and work each regular problem. Next to the regular problem, work an estimation problem with the numbers rounded to the nearest hundred.

P6						
a.	2 4 1	▇		b.	3 9 1	▇
	+ 5 6 3	+ ▇			+ 5 2 2	+ ▇

Part 7 Copy each item and write the sign >, <, or =.

P7						
a. $\frac{5}{7}$ ▇ $\frac{4}{4}$		b. $\frac{3}{3}$ ▇ $\frac{9}{9}$		c. $\frac{5}{5}$ ▇ $\frac{8}{9}$

Part 8 Write the fraction for each number line.

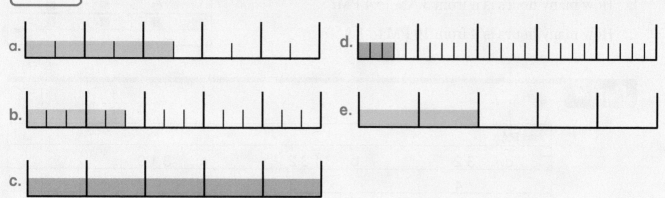

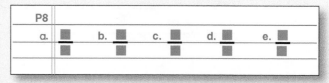

Lesson 70

Part 9 | Copy each item and write the sign >, <, or =.

P9	
a.	3×3 ■ $5 + 4$ b. $9 - 1$ ■ 2×5

Part 10 | For each item, write the column problem to find the missing number.

a. $50 - $ ■ $= 20$ b. ■ $- 20 = 10$ c. $30 + $ ■ $= 50$

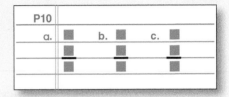

Part 11 | Write the numbers for counting by 4s to 40.

P11	
a.	4 ■ ■ ■ ■ ■ ■ ■ ■ 40

Part 12 | Write the time for each clock.

P12	
a.	■ : ■ b. ■ : ■ c. ■ : ■ d. ■ : ■

a.

b.

c.

d.
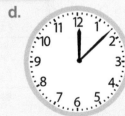

Lesson 71

Part 1

a. $4\overset{31}{\longrightarrow}\blacksquare$ b. $3\overset{52}{\longrightarrow}\blacksquare$ c. $12\overset{4}{\longrightarrow}\blacksquare$

P1			
a. ■	b. ■	c. ■	
■	■	■	
━	━	━	
■	■	■	

Part 2

a. Each pencil cost 4 cents.

b. There are 5 girls on each team.

c. There are 8 windows in each room.

d. Each week has 7 days.

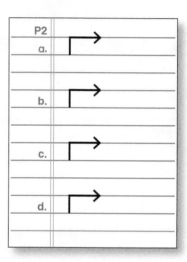

P2	
a.	→
b.	→
c.	→
d.	→

Part 3

Living Room	👤
Dining Room	👤 👤 👤 👤
Kitchen	👤 👤
Bedroom	👤 👤 👤

👤 = 4 people

P3	
a.	b.
c.	d.

a. How many people were in the kitchen?

b. How many people were in the bedroom?

c. Which room had the most people?

d. Which room had 4 people?

Lesson 71

Part 4

a. J $\xrightarrow{2}$ 10   b. 5 $\xrightarrow{11}$ M   c. 3 $\xrightarrow{C}$ 15

d. P $\xrightarrow{5}$ 40   e. 2 $\xrightarrow{20}$ G   f. 3 $\xrightarrow{12}$ R

P4		
a.	b.	c.
d.	e.	f.

Part 5

a. How many hours is it from 1 AM to 10 PM?

b. How many hours is it from 9 AM to 7 PM?

c. How many hours is it from 8 PM to 3 AM?

d. How many hours is it from 5 AM to 6 PM?

P5			
a. ■	b. ■	c. ■	d. ■
■	■	■	■
■	■	■	■

Independent Work

Part 6 | Work each problem. Write the unit name.

a. The bookstore had 1500 books at the beginning of the day.
The bookstore sold a lot of books during the day.
At the end of the day, the bookstore had 1150 books.
How many books did the bookstore sell?

b. There were clean plates and dirty plates in the kitchen.
There were 45 plates in all.
19 plates were clean.
How many plates were dirty?

c. The train went from New York to Chicago.
The train started out with 158 people in New York.
124 people got on the train during the trip.
How many people were on the train when it arrived in
Chicago?

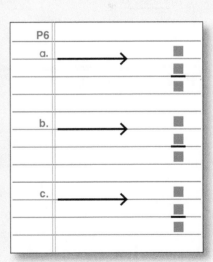

Lesson 71

Part 7 | Copy and work each problem.

P7

a.	412	b. 120	c. 510
	× 3	× 4	× 3

Part 8 | Write the place-value addition fact for each number.

a. 4912   b. 2351   c. 1111

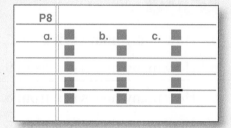

Part 9 | Copy and work each problem.

a. 230   b. 400   c. 1500
  − 85   − 70   − 950

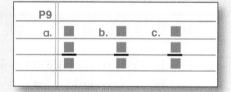

Part 10 | Write the time for each clock.

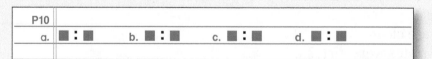

a.

b.

c.

d.

Lesson 71

Part 11 Copy and work each regular problem. Next to the regular problem, work an estimation problem with the numbers rounded to the nearest hundred.

P11				
a.	1 9 5	▇	b. 3 0 7	▇
	+ 4 1 6	+ ▇	+ 4 9 4	+ ▇

Part 12 Copy each item and write the sign >, <, or =.

P12			
a.	$\frac{4}{4}$ ▇ $\frac{7}{9}$	b. $\frac{5}{5}$ ▇ $\frac{3}{2}$	c. $\frac{4}{3}$ ▇ $\frac{8}{8}$

Part 13

a. Find the perimeter of this rectangle.

b. Find the area of this rectangle.

10 cm

5 cm

P13	
a.	▇ + ▇ = ▇
	▇ + ▇ = ▇
	▇
b.	▇ × ▇ = ▇

Lesson 72

Part 1

November	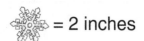
December	
January	
February	

 = 2 inches

P1	
a.	b.
c.	d.

a. How many inches of snow fell in January?

b. How many inches of snow fell in November?

c. Which two months had the same amount of snow?

d. Which month had 10 inches of snow?

Part 2

a. Each dime is worth 10 cents.
Tom had 8 dimes.
How many cents did he have?

b. Every chair has 4 legs.
There are 12 legs.
How many chairs are there?

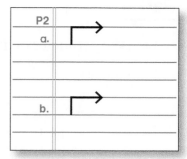

Part 3

a. How many hours is it from 7 AM to 6 PM?

b. How many hours is it from 9 PM to 4 AM?

c. How many hours is it from 10 AM to 1 PM?

d. How many hours is it from 9 PM to 7 AM?

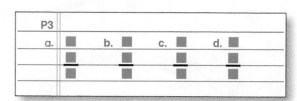

Lesson 72

Independent Work

Part 4 Work each problem. Write the unit name.

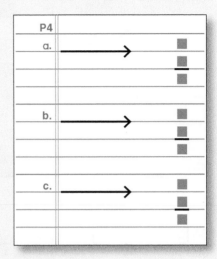

a. Gary has $9.50. Bob has $2.75.
   How much more money does Gary have than Bob has?

b. There are 24 red pens in the can.
   There are 45 pens that are not red in the can.
   How many pens are there altogether in the can?

c. The cow weighed 95 pounds less than a horse.
   The cow weighed 345 pounds.
   How many pounds did the horse weigh?

Part 5 Copy and work each problem.

P5			
a.	2 0 2	b. 1 3 0	c. 4 1 2
	× 4	× 3	× 2

Part 6 Write the place-value addition fact for each number.

a. 2416   b. 5555   c. 3142

Lesson 72

Part 7 Copy each problem and write the answer.

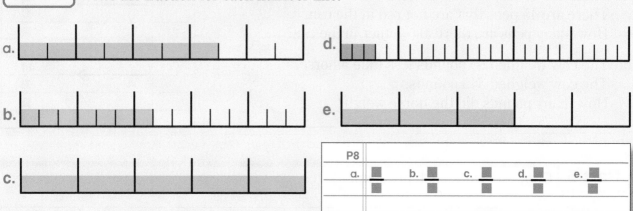

P7	
a. $\dfrac{3}{4} - \dfrac{2}{4} =$ ___ b. $\dfrac{7}{9} + \dfrac{1}{9} =$ ___ c. $\dfrac{7}{9} - \dfrac{1}{9} =$ ___	

Part 8 Write the fraction for each number line.

a.

b.

c.

d.

e.

P8					
a.	b.	c.	d.	e.

Lesson 72

Part 9 Copy and work each regular problem. Next to the regular problem, work an estimation problem with the numbers rounded to the nearest hundred.

P9				
a.	194	▇	b. 387	▇
	+407	+ ▇	+495	+ ▇

Part 10 Copy and work each problem.

a. 420
   − 95

b. 4400
   − 990

c. 530
   − 155

P10			
a. ▇	b. ▇	c. ▇
▇	▇	▇
▇	▇	▇

Part 11 Write the numbers for counting by 4s to 40.

P11	
a.	4 ▇ ▇ ▇ ▇ ▇ ▇ ▇ 40

Lesson 73

Part 1

Monday	🚗 🚗 🚗
Tuesday	🚗
Wednesday	🚗 🚗
Thursday	🚗 🚗 🚗 🚗 🚗 🚗 🚗 🚗

🚗 = 5 cars

a. How many cars were in the lot on Monday?

b. How many cars were in the lot on Thursday?

c. On which day were there 10 cars in the lot?

d. How many more cars were in the lot on Monday than on Tuesday?

e. How many more cars were in the lot on Wednesday than on Tuesday?

P1		
a.	b.	c.
d.	e.	

Part 2

a. There were 4 trucks.
Each truck had 12 wheels.
How many wheels were there?

b. Each room had 5 tables.
There were 11 rooms.
How many tables were there?

c. Linda ran 4 miles every day.
She ran 12 miles.
How many days did she run?

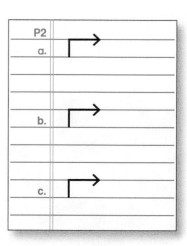

Lesson 73

Part 3

a. $8\overline{)8}$ b. $8\overline{)0}$ c. $4\overline{)0}$

d. $6\overline{)6}$ e. $9\overline{)0}$

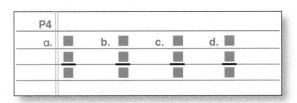

P3					
	a.	b.	c.	d.	e.

Part 4

a. How many hours is it from 5 AM to 10 PM?

b. How many hours is it from 3 PM to 3 AM?

c. How many hours is it from 6 AM to 9 PM?

d. How many hours is it from 9 PM to 6 AM?

P4			
a. ■	b. ■	c. ■	d. ■
■	■	■	■
■	■	■	■

Independent Work

Part 5 | Work each problem. Write the unit name.

a. The red ribbon is 135 centimeters long.
The blue ribbon is 254 centimeters long.
How much longer is the blue ribbon than the red ribbon?

b. The big box weighed 130 pounds.
The big box weighed 28 pounds more than the small box.
How many pounds did the small box weigh?

c. In the morning, Jerry read 74 pages of a book.
During the afternoon, Jerry read some more pages.
By the end of the afternoon, Jerry had read 125 pages.
How many pages did Jerry read during the afternoon?

d. The flower shop had roses.
During the day, the shop sold 125 roses.
At the end of the day, the shop still had 126 roses.
How many roses did the shop have at the beginning
of the day?

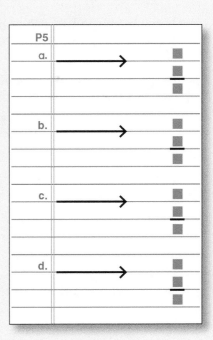

Lesson 73

Part 6 Copy and work each problem.

P6			
a.	4 0	b. 2 0 3	c. 5 1 0
	× 2	× 3	× 2

Part 7

$2.94 $9.05 $2.17 $6.82 $4.18

a. Jerry buys the hat, the book, and the pen. **About** how much does he spend?

b. Jerry buys the hat, the book, and the pen. **Exactly** how much does he spend?

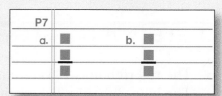

Part 8 Write the place-value addition fact for each number.

a. 5326   b. 8414   c. 9425

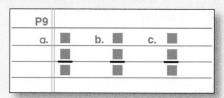

Part 9 Write and work the problems to find the missing number.

a. 90 − ■ = 40   b. ■ − 60 = 10   c. ■ − 5 = 4

Lesson 73

Part 10 Write the fraction for each picture.

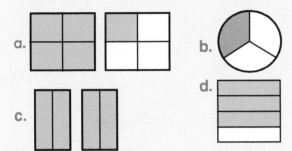

a.

b.

c.

d.

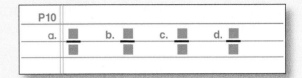

P10				
a.	b.	c.	d.

Part 11 Write the numbers for counting by 4s to 40 and 3s to 30.

P11	
a.	4 ▪ ▪ ▪ ▪ ▪ ▪ ▪ ▪ ▪ 40
b.	3 ▪ ▪ ▪ ▪ ▪ ▪ ▪ ▪ ▪ 30

Lesson 74

Part 1

a. Joe earned 5 dollars every hour he worked.
He earned 30 dollars.
How many hours did he work?

b. Mrs. Green made pies for 5 weeks.
She made 10 pies every week.
How many pies did she make in all?

c. Each row has 5 chairs.
There are 20 rows.
How many chairs are there?

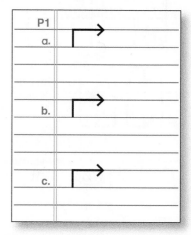

Part 2

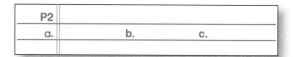

Teachers	
Firefighters	
Lawyers	
Doctors	

= 10 women

a. How many women are teachers?

b. How many women are firefighters?

c. How many more women are teachers than firefighters?

Independent Work

Part 3 For each problem, figure out the hours before 12. Add those hours to the hours after 12.

a. How many hours is it from 9 PM to 7 AM?

b. How many hours is it from 2 AM to 4 PM?

c. How many hours is it from 8 PM to 6 AM?

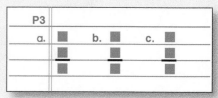

Lesson 74

Part 4 Work each problem. Write the unit name.

a. Kevin has 92 more marbles than Raymond has.
Raymond has 48 marbles.
How many marbles does Kevin have?

b. 425 people are going to a school to learn how to fly planes.
155 of the people are women.
How many men are going to the school?

c. Alice started out with some money.
She spent $5.45 of the money she earned on a new book.
Now she has $9.55 left.
How much money did she start with?

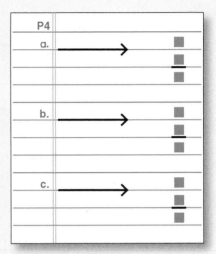

Part 5 Copy and work each problem.

P5			
a. 5 3	b. 4 2 1	c. 3 1 0
× 2	× 3	× 3

Part 6

a. ⟨ $6.85 ⟩   b. ⟨ $6.05 ⟩   c. ⟨ $1.62 ⟩   d. ⟨ $1.12 ⟩   e. ⟨ $2.64 ⟩

1. Marcos bought two of these items. One item was item C.
He spent about 9 dollars. Which two items did he buy?

2. Bob bought two items. One item was item B.
He spent about 8 dollars. Which two items did he buy?

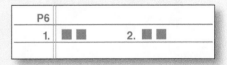

Part 7 Write the answer to each place-value problem.

a. 4000 + 900 + 90 + 9   b. 1000 + 100 + 10 + 2

c. 2000 + 600 + 60 + 1   d. 1000 + 900 + 90 + 9

P7			
a.	b.	c.	d.

Lesson 74

Part 8 Copy each item and write the sign >, <, or =.

P8			
a.	4 × 4 ■ 9 + 9	b. 10 × 2 ■ 5 × 4	c. 3 × 4 ■ 10 + 3

Part 9 Write the answer to each problem.

a. 4$\overline{)20}$ b. 4$\overline{)12}$ c. 4$\overline{)4}$ d. 4$\overline{)16}$ e. 4$\overline{)8}$

f. 5$\overline{)30}$ g. 5$\overline{)40}$ h. 5$\overline{)20}$ i. 5$\overline{)45}$ j. 5$\overline{)35}$

P9				
a.	b.	c.	d.	e.
f.	g.	h.	i.	j.

Part 10 Write the answer to each problem.

a. 42 + 7   b. 74 + 5   c. 32 + 6

d. 73 + 5   e. 24 + 4

P10				
a.	b.	c.	d.	e.

Connecting Math Concepts

Lesson

a. Millie ate 2 apples every day.
   She ate apples for 13 days.
   How many apples did she eat?

b. Mrs. James worked in her garden for 20 hours.
   She worked for 4 hours each day.
   How many days did she work in the garden?

c. Each room had 5 windows.
   There are 30 windows.
   How many rooms are there?

d. José has 12 shirts.
   Each shirt has 4 buttons.
   How many buttons are there on all the shirts?

P1	
a.	→
b.	→
c.	→
d.	→

Independent Work

Part 2 Write the answer to each question.

This graph shows how many days it snowed during four months.

P2	
a.	b.
c.	d.

December	❄ ❄ ❄
January	❄ ❄ ❄ ❄ ❄
February	❄ ❄
March	❄

❄ = 3 days

a. How many days did it snow in December?

b. How many days did it snow in March?

c. How many more days did it snow in December than in March?

d. How many days did it snow in January?

Lesson 75

Part 3 | Work each problem. Write the unit name.

a. Gary started out with $19.50.
   He bought something that cost $2.75.
   How much money did he end up with?

b. There were 95 chickens in the barn.
   There were 45 more turkeys than chickens in the barn.
   How many turkeys were in the barn?

c. Roberto worked 134 hours in April and 99 hours in May.
   How many more hours did Roberto work in April than
   in May?

d. The store had lots of apples.
   The store sold 482 apples.
   Now the store has 385 apples.
   How many apples did the store start out with?

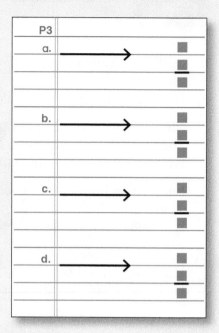

Part 4 | For each problem, figure out the hours before 12. Add those hours to the hours after 12.

a. How many hours is it from 7 AM to 4 PM?

b. How many hours is it from 10 PM to 6 AM?

c. How many hours is it from 4 PM to 4 AM?

d. How many hours is it from 7 AM to 5 PM?

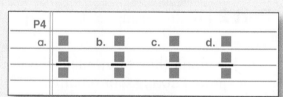

Part 5 | Write the answer to each place-value problem.

a. 3000 + 200 + 10 + 4   b. 1000 + 900 + 10 + 2

c. 5000 + 500 + 50 + 5   d. 3000 + 100 + 40 + 9

Part 6 | Write the answer to each problem.

a. $5\overline{)35}$ b. $5\overline{)25}$ c. $5\overline{)30}$ d. $5\overline{)5}$ e. $5\overline{)35}$

f. $2\overline{)16}$ g. $2\overline{)14}$ h. $2\overline{)2}$ i. $4\overline{)12}$ j. $3\overline{)12}$

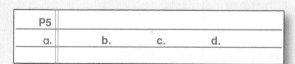

Connecting Math Concepts

Part 7 Copy and work each regular problem.
Next to the regular problem, work an estimation problem.

P7				
a.	7 4	▮	b. 5 8	▮
	+ 2 8	+ ▮	+ 3 1	+ ▮

Part 8 Work each problem.

a. How much more is 900 than 350?

b. How much less is 200 than 600?

P8	
a. →	▮ ▮ ▮
b. →	▮ ▮ ▮

Lesson 76

Part 1

a. There were 12 fleas on each dog.
   There were 4 dogs.
   How many fleas were there?

b. Each classroom has 22 students.
   There are 3 classrooms.
   How many students are there?

c. Jamal ran 5 miles every day.
   Jamal ran 35 miles.
   How many days did he run?

d. Every hour the baker made 4 pies.
   He made 20 pies.
   How many hours did he make pies?

Part 2

a. 123¢   b. 9¢   c. 912¢

d. 600¢   e. 20¢   f. 2¢

Independent Work

Part 3 Work each problem. Write the unit name.

a. There are 125 plates.
   95 of the plates are clean.
   How many of the plates are dirty?

b. Sam had 11 fewer friends than Alex had.
   Sam had 29 friends.
   How many friends did Alex have?

c. The boy worked 296 math problems.
   145 of the problems were addition problems.
   The rest were subtraction problems
   How many subtraction problems did the boy work?

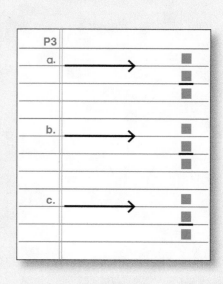

Lesson 76

Part 4 Write the place-value addition fact for each number.

a. 3212   b. 1943   c. 1111

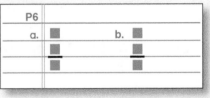

Part 5 Copy and work each problem.

P5			
a.	5 0 × 9	b. 4 3 1 × 3	c. 3 0 1 × 9

Part 6 For each problem, figure out the hours before 12. Add those hours to the hours after 12.

a. How many hours is it from 9 AM to 5 PM?

b. How many hours is it from 10 PM to 7 AM?

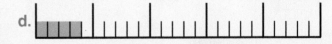

Part 7 Write the fraction for each number line.

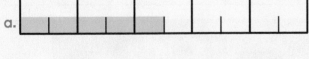

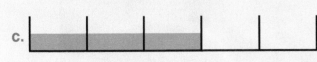

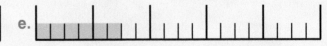

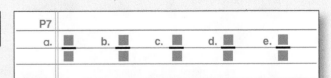

Lesson 76

Part 8 Write **Cu** for cube, **Py** for pyramid, **Sp** for sphere, **H** for hexagon, **P** for pentagon, **R** for rectangle, **S** for square, **T** for triangle.

 1.
 2.
 3.
 4.

 5.
 6.
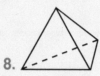 7.
8.

P8			
1.	2.	3.	4.
5.	6.	7.	8.

Part 9 Copy and work each problem.

a. $\begin{array}{r} \$5.00 \\ -1.30 \\ \hline \end{array}$ b. $\begin{array}{r} \$25.00 \\ -4.50 \\ \hline \end{array}$

P9		
a. ■		b. ■
■		■
■		■

Part 10 Write the answer to each problem.

a. $2\overline{)16}$ b. $5\overline{)0}$ c. $5\overline{)30}$ d. $2\overline{)14}$ e. $4\overline{)4}$

f. $5\overline{)35}$ g. $3\overline{)12}$ h. $2\overline{)2}$ i. $2\overline{)0}$ j. $5\overline{)40}$

P10				
a.	b.	c.	d.	e.
f.	g.	h.	i.	j.

Lesson 77

Part 1

a. Each room has 5 windows.
There are 20 rooms.
How many windows are there?

b. Jill read a book for 7 days.
She read 20 pages every day.
How many pages did she read?

c. Jada drew 4 pictures every day.
Jada drew 12 pictures.
How many days did she draw pictures?

d. Each bag has 5 apples.
There are 30 apples.
How many bags of apples are there?

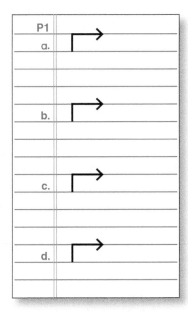

Part 2

P2				
1.	2.	3.	4.	5.

1. Which park has 5 benches?

2. Which park has the most benches?

3. Which park has the fewest benches?

4. How many benches does Park A have?

5. How many benches does Park D have?

Part 3 Copy each item and write the sign >, <, or =.

P3			
a. $\dfrac{5}{3}$ ■ $\dfrac{7}{7}$	b. $\dfrac{3}{2}$ ■ $\dfrac{5}{6}$	c. $\dfrac{9}{10}$ ■ $\dfrac{5}{4}$
d. $\dfrac{4}{4}$ ■ $\dfrac{7}{9}$	e. $\dfrac{4}{5}$ ■ $\dfrac{7}{5}$	f. $\dfrac{5}{9}$ ■ $\dfrac{3}{2}$

Connecting Math Concepts

Lesson 77 **175**

Independent Work

Part 4 Write the cents amount with a dollar sign and a dot.

a. 106¢   b. 3¢   c. 540¢

d. 89¢   e. 7¢   f. 611¢

P4			
a.		b.	c.
d.		e.	f.

Part 5 Work each problem. Write the unit name.

a. The bull weighs 145 pounds more than the horse.
   The bull weighs 642 pounds.
   How many pounds does the horse weigh?

b. The train started out with lots of passengers.
   45 passengers got off the train during the trip.
   The train had 157 passengers when it finished the trip.
   How many passengers did the train start with?

c. The temperature on Monday was 19 degrees higher
   than on Tuesday.
   The temperature was 54 degrees on Monday.
   How many degrees was the temperature on Tuesday?

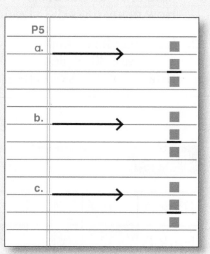

Part 6 Copy and work each problem.

P6			
a.	2 1 0 × 5	b. 4 3 0 × 3	c. 2 1 0 × 9

Part 7 For each problem, figure out the hours before 12. Add those hours to the hours after 12.

a. How many hours is it from 8 PM to 1 AM?

b. How many hours is it from 9 AM to 4 PM?

c. How many hours is it from 5 PM to 3 AM?

P7			
a. ■	b. ■	c. ■	

Lesson 77

Part 8 Copy and work each regular problem. Next to the regular problem, work an estimation problem with the numbers rounded to the nearest hundred.

P8				
a.	2 9 4	▆	b. 1 8 4	▆
	+1 8 6	+ ▆	+3 1 7	+ ▆

Part 9 Copy and work each problem.

a. $42.80
   −11.90

b. $30.60
   − 1.50

c. $8.50
   +1.50

P9			
a.	▪	b. ▪	c. ▪
	▬	▬	▬
	▪	▪	▪

Part 10 Write **Cu** for cube, **Py** for pyramid, **Sp** for sphere, **H** for hexagon, **P** for pentagon, **R** for rectangle, **S** for square, **T** for triangle.

1.   2.   3.   4.

5. □ 6. ⬭ 7.   8. ⬛

P10				
1.	2.	3.	4.
5.	6.	7.	8.

Part 11 Write the place-value addition fact for each number.

a. 3421   b. 1816

P11		
a.	▪	b. ▪
	▪	▪
	▪	▪
	▬	▬
	▪	▪

Part 12 Write the fraction for each number line.

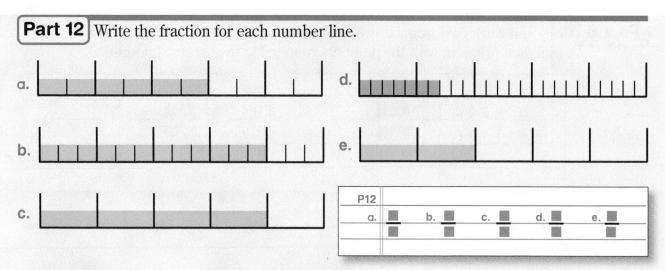

Lesson 78

Part 1

P1			
a.	b.	c.	
d.	e.	f.	

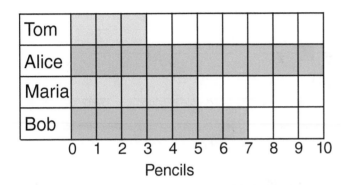

a. How many pencils does Tom have?

b. How many pencils does Alice have?

c. Which student has the most pencils?

d. Which student has the fewest pencils?

e. Which student has 7 pencils?

f. Which student has more than 8 pencils?

Part 2

a. Tam eats 4 apples every week.
She eats apples for 12 weeks.
How many apples does she eat?

b. There are 20 trees.
There are 5 trees on each street.
How many streets are there?

c. Jerry works for 5 hours.
Every hour has 60 minutes.
How many minutes does Jerry work?

d. Each street has 2 houses.
There are 12 houses.
How many streets are there?

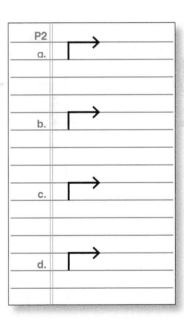

Lesson 78

Independent Work

Part 3 Copy each item and write the sign >, <, or =.

P3			
a. $\dfrac{5}{6}$ ■ $\dfrac{2}{2}$	b. $\dfrac{5}{3}$ ■ $\dfrac{7}{9}$	c. $\dfrac{9}{10}$ ■ $\dfrac{4}{3}$
d. $\dfrac{3}{3}$ ■ $\dfrac{6}{5}$	e. $\dfrac{3}{7}$ ■ $\dfrac{9}{7}$	f. $\dfrac{6}{4}$ ■ $\dfrac{8}{9}$

Part 4 Write the answer to each question.

This graph shows how many houses are on each street.

James Street	🏠
Adams Street	🏠 🏠 🏠 🏠 🏠
High Street	🏠 🏠
River Street	🏠 🏠 🏠

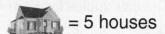 = 5 houses

a. How many houses are on River Street?

b. Which street has 5 houses?

c. Which street has 5 more houses than High Street?

d. Which street has fewer than 10 houses?

P4	
a.	b.
c.	d.

Part 5 Copy and work each problem.

P5			
a. $\begin{array}{r} 402 \\ \times\ \ 3 \\ \hline \end{array}$	b. $\begin{array}{r} 104 \\ \times\ \ 2 \\ \hline \end{array}$	c. $\begin{array}{r} 510 \\ \times\ \ 9 \\ \hline \end{array}$

Connecting Math Concepts

Lesson 80

Part 4 Write the cents amount with a dollar sign and a dot.

a. 211¢   b. 72¢   c. 100¢

d. 6¢   e. 817¢   f. 9¢

P4		
a.	b.	c.
d.	e.	f.

Part 5 Copy each item and write the sign >, <, or =.

P5			
a. $\dfrac{5}{3}$ ■ $\dfrac{6}{10}$	b. $\dfrac{4}{8}$ ■ $\dfrac{8}{4}$	c. $\dfrac{3}{2}$ ■ $\dfrac{6}{9}$
d. $\dfrac{4}{7}$ ■ $\dfrac{5}{4}$	e. $\dfrac{4}{3}$ ■ $\dfrac{3}{6}$	f. $\dfrac{4}{4}$ ■ $\dfrac{6}{6}$

Part 6 For each problem, figure out the hours before 12. Add those hours to the hours after 12.

a. How many hours is it from 9 AM to 5 PM?

b. How many hours is it from 8 PM to 7 AM?

c. How many hours is it from 9 AM to 9 PM?

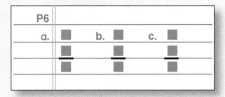

Part 7 For each item, work the column problem to find the missing number.

a. 1200 + ■ = 2000   b. 500 + ■ = 750

Lesson 80

Part 8 Work each problem. Write the unit name.

a. Tom is 46 centimeters shorter than Bob.
Tom is 135 centimeters tall.
How many centimeters tall is Bob?

b. There are 65 more apples than oranges.
There are 120 oranges.
How many apples are there?

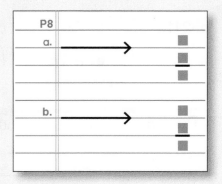

Part 9 Write the place-value addition fact for each number.

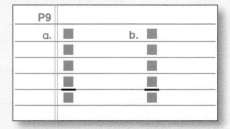

a. 1516   b. 2434

Part 10 Write the answer to each problem.

a. 2⟌1 6   b. 4⟌1 2   c. 2⟌1 4   d. 4⟌1 6   e. 2⟌1 6

f. 5⟌3 0   g. 5⟌4 0   h. 5⟌2 0   i. 5⟌4 5   j. 5⟌3 5

P10				
a.	b.	c.	d.	e.
f.	g.	h.	i.	j.

Lesson 81

Part 1

Lilies									
Roses									
Violets									
Pansies									

0   5   10   15   20   25   30   35   40   45

Flowers in a Garden

P1				
a. ■	b. ■	c. ■	d. ■

a. How many more pansies than lilies are in the garden?

b. How many more violets than roses are in the garden?

c. How many more pansies than roses are in the garden?

d. How many more violets than lilies are in the garden?

Part 2

P2			
a.	b.	c.
d.	e.	f.

a. The boat was 16 feet shorter than the truck.
The boat was 36 feet long.
How long was the truck?

b. There are 9 windows on each floor of a building. There are 14 floors.
How many windows are in the building?

c. Bill had $65. He spent $47.
How many dollars did he end up with?

d. Every team had 5 players.
There were 120 players.
How many teams were there?

e. Each box weighed 7 pounds.
There were 20 boxes.
How much did they weigh altogether?

f. Gary is 14 years older than his sister.
His sister is 43 years old.
How old is Gary?

Lesson

Part 3

	Monday	Tuesday	Wednesday
Jan	$40	$40	$40
Pete	$66	$0	$81
Ms. Brown	$21	$100	$100

a. How much did Ms. Brown earn on the three days?

b. How much did Jan earn on the three days?

c. How much did Pete earn on the three days?

Independent Work

Part 4 Copy and work each problem.

P4				
a.	43	b.	510	c.	24	d.	67

a.
```
    43
  ×  5
```
b.
```
   510
  ×  2
```
c.
```
    24
  ×  9
```
d.
```
    67
  ×  2
```

Part 5 Write the fraction for each picture.

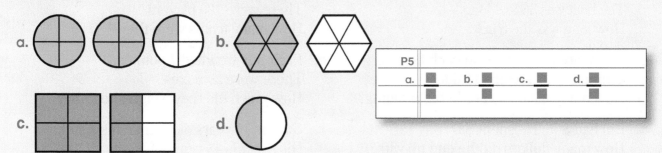

a.

b.

c.

d.

Lesson 81

Part 6 Work each problem. Write the unit name.

   a. There are 5 windows in each room.
      There are 32 rooms.
      How many windows are there?

   b. Every car has 4 tires.
      There are 20 tires.
      How many cars are there?

   c. Every shirt has 8 buttons.
      There are 11 shirts.
      How many buttons are there?

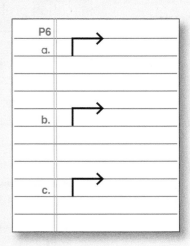

Part 7 Copy each problem and write the answer.

P7

  a. $\dfrac{5}{9} - \dfrac{2}{9} = $ ___   b. $\dfrac{3}{5} + \dfrac{1}{5} = $ ___   c. $\dfrac{4}{10} - \dfrac{3}{10} = $ ___

Part 8 Write the cents amount with a dollar sign and a dot.

a. 300¢   b. 9¢   c. 45¢   d. 620¢

P8			
a.	b.	c.	d.

Part 9 Work each problem. Write the unit name.

a. The month had 31 days.
   It rained on 14 days.
   How many days did it not rain?

b. There are 25 new cars and 155 old cars in the parking lot.
   How many cars in all are in the parking lot?

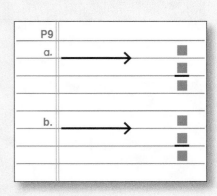

Lesson 81

Part 10 Copy and work each regular problem. Next to the regular problem, work an estimation problem with the numbers rounded to the nearest hundred.

P10				
a.	384	▇	b.	415 ▇
	+217	+ ▇		+686 + ▇

Part 11 Copy each item and write the sign >, <, or =.

P11	
a.	5 × 3 ▇ 10 + 5 b. 9 − 1 ▇ 6 + 3

Lesson 82

Part 1

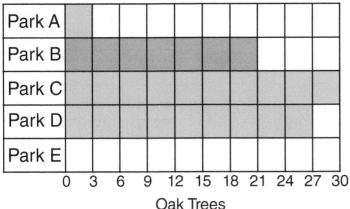

| | 0 | 3 | 6 | 9 | 12 | 15 | 18 | 21 | 24 | 27 | 30 |
Oak Trees

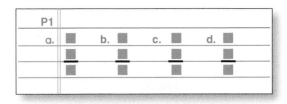

a. How many more oak trees were in Park C than Park A?

b. How many more oak trees were in Park D than Park B?

c. How many more oak trees were in Park B than Park E?

d. How many more oak trees were in Park C than Park B?

Part 2

a.   b.   c.   d.   e.   f.

1. Write the letter of each shape that is a triangle.

2. Write the letter of each shape that is a rectangle.

3. Write the letter of each shape that is a hexagon.

4. Write the letter of each shape that is a quadrilateral.

5. Write the letter of each shape that is a square.

P2		
1.	2.	3.
4.	5.	

Lesson 82

Part 3

a. Fred weighed 24 pounds less than Jim.
   Jim weighed 128 pounds.
   How many pounds did Fred weigh?

b. There were 106 vehicles.
   12 of the vehicles were red.
   How many vehicles were not red?

c. There were 5 people in every car.
   There were 45 people in all.
   How many cars were there?

d. Each can weighed 3 pounds.
   How many pounds did 15 cans weigh?

Part 4

	Monday	Tuesday	Wednesday
Jan	$40	$40	$40
Pete	$66	$0	$81
Ms. Brown	$21	$100	$100

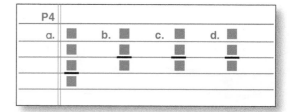

a. How much did the people earn on Monday?

b. How much money did Jan and Pete earn on Wednesday?

c. How much more money did Pete earn on Wednesday than he earned on Monday?

d. How much more money did Pete earn on Wednesday than Jan earned on Wednesday?

Independent Work

Part 5 For each problem, figure out the hours before 12. Add those hours to the hours after 12.

a. How many hours is it from 4 PM to 2 AM?

b. How many hours is it from 8 AM to 5 PM?

c. How many hours is it from 10 PM to 6 AM?

Lesson

Part 6 | Copy each item and write the sign >, <, or =.

P6			
a. $\dfrac{7}{4}$ ■ $\dfrac{7}{10}$	b. $\dfrac{4}{5}$ ■ $\dfrac{3}{3}$	c. $\dfrac{4}{4}$ ■ $\dfrac{8}{8}$
d. $\dfrac{6}{7}$ ■ $\dfrac{7}{6}$	e. $\dfrac{4}{3}$ ■ $\dfrac{4}{9}$	f. $\dfrac{2}{2}$ ■ $\dfrac{3}{2}$

Part 7 | Write the answer to each question.

This graph shows how many pages of a book each student read.

P7			
a.	b.	c.	d.

Jerry	📖 📖 📖 📖
Alex	📖 📖
Susan	📖 📖 📖 📖 📖
Tammy	📖 📖 📖 📖

📖 = 5 pages

a. What student read 10 pages?

b. How many pages did Susan read?

c. How many more pages did Tammy read than Alex read?

d. How many more pages did Susan read than Alex read?

Part 8 | Copy and work each problem.

P8			
a. $\begin{array}{r} 53 \\ \times\ 4 \\ \hline \end{array}$	b. $\begin{array}{r} 203 \\ \times\ 3 \\ \hline \end{array}$	c. $\begin{array}{r} 420 \\ \times\ 4 \\ \hline \end{array}$	d. $\begin{array}{r} 19 \\ \times\ 2 \\ \hline \end{array}$

Part 9

a. Find the perimeter of this rectangle.

b. Find the area of this rectangle.

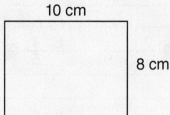

10 cm

8 cm

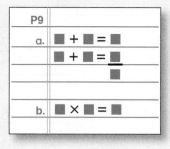

P9	
a.	■ + ■ = ■
	■ + ■ = ■
	___
	■
b.	■ × ■ = ■

Part 10 Write the time for each clock.

P10	
a. ■ : ■	b. ■ : ■ c. ■ : ■ d. ■ : ■

a.

b.

c.

d.

Part 11 Write the fraction for each number line.

a.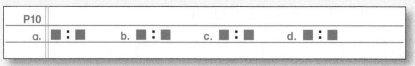

d.

b.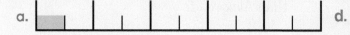

e.

c.

f.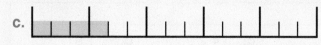

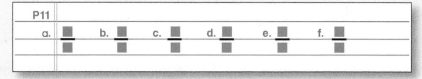

P11	
a. ■ b. ■ c. ■ d. ■ e. ■ f. ■	
■ ■ ■ ■ ■ ■	

Connecting Math Concepts

Lesson 83

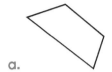

Part 1

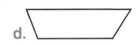

a.　　b.　　c.　　d.

e.　　f.　　g.　　h.

1. Write the letter of each shape that is a hexagon.

2. Write the letter of each shape that is a quadrilateral.

3. Write the letter of each shape that is a pentagon.

4. Write the letter of each shape that is a rectangle.

5. Write the letter of each shape that is a square.

6. Write the letter of each shape that is a triangle.

P1			
1.	2.	3.
4.	5.	6.

Part 2

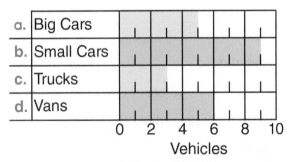

P2				
a.	b.	c.	d.	

Lesson 83

Part 3

a. Roberto had 25 more marbles than Jim had.
Roberto had 44 marbles.
How many marbles did Jim have?

b. Each day Ming reads 5 pages.
She reads for 15 days.
How many pages does she read?

c. There are 4 cookies in every package.
There are 80 cookies.
How many packages are there?

d. The tree was 15 feet taller than the house.
The tree was 38 feet high.
How many feet high was the house?

P3	
a.	
b.	
c.	
d.	

Part 4

	Monday	Tuesday	Wednesday
Henry	$165	$175	$105
Alice	$68	$86	$81
Mr. Black	$60	$80	$90

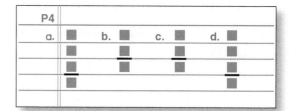

a. How much money did the people earn on Wednesday?

b. How much more money did Alice earn on Tuesday than she earned on Wednesday?

c. How much less did Mr. Black earn on Monday than he earned on Tuesday?

d. How much money did Henry earn on the three days?

Independent Work

Part 5 Copy and work each problem.

P5				
a.	47 × 5	b. 13 × 9	c. 401 × 3	d. 24 × 3

Lesson 83

Part 6 Write the cents amount with a dollar sign and a dot.

a. 200¢   b. 20¢   c. 2¢

P6			
a.		b.	c.

Part 7 Copy each item and write the sign >, <, or =.

P7			
a. $\dfrac{4}{4}$ ■ $\dfrac{3}{2}$	b. $\dfrac{4}{3}$ ■ $\dfrac{6}{9}$	c. $\dfrac{5}{5}$ ■ $\dfrac{9}{9}$	

Part 8 Write the place-value addition fact for each number.

a. 5555   b. 3333

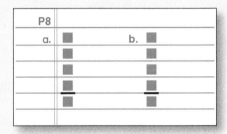

Part 9 Copy and work each problem.

a. $5.20
   −1.05

b. $9.00
   −1.40

Lesson 83

Part 10 Write the answer to each question.

This graph shows how many cars Tina fixed on different days.		

Monday	🚗 🚗 🚗
Tuesday	🚗 🚗 🚗 🚗 🚗
Wednesday	🚗
Thursday	🚗 🚗 🚗 🚗
Friday	🚗 🚗 🚗

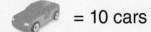 = 10 cars

P10		
a.	b.	c.

a. How many cars did Tina fix on Friday?

b. On which day did Tina fix 10 cars?

c. How many more cars did Tina fix on Thursday than on Wednesday?

Part 11 Work each problem. Write the unit name.

a. Tom is 126 centimeters tall.
   Tom is 46 centimeters shorter than Bob.
   How many centimeters tall is Bob?

b. Jerry had $9.00. He earned some money.
   He ended up with $14.30. How much money did he earn?

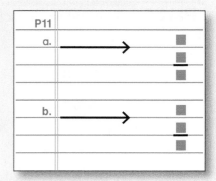

Part 12

a. Find the perimeter of this rectangle.

b. Find the area of this rectangle.

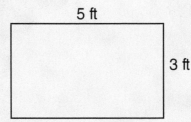

5 ft

3 ft

Lesson 84

Part 1

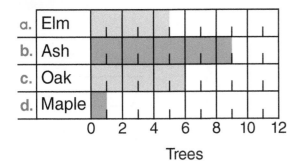

	P1	
	a.	b. c. d.

Part 2

a. Jerry has 10 shirts.
   There are 9 buttons on each shirt.
   How many buttons are there?

b. Ann is 24 years younger than Mary.
   Ann is 42 years old.
   How old is Mary?

c. The store started the day with 95 computers.
   The store sold 45 computers during the day.
   How many computers did the store end up with?

d. Every day has 24 hours.
   How many hours are in 3 days?

P2
a.
b.
c.
d.

Part 3

	Monday	Tuesday	Wednesday
Henry	$165	$175	$145
Alice	$68	$86	$81
Mr. Black	$60	$80	$90

P3			
a. ■	b. ■	c. ■	
■	■	■	
■	■	■	

a. How much less money did Henry earn on Monday than on Tuesday?

b. On Wednesday, how much more money did Mr. Black earn than Alice earned?

c. On Monday, how much more money did Henry earn than Alice earned?

Lesson

Part 4 | Write the answer to each item.

a.   b.   c.   d.

e. 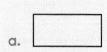   f.   g.   h.

1. Write the letter of each shape that is a hexagon.

2. Write the letter of each shape that is a triangle.

3. Write the letter of each shape that is a square.

4. Write the letter of each shape that is a rectangle.

5. Write the letter of each shape that is a quadrilateral.

6. Write the letter of each shape that is a pentagon.

P4		
1.	2.	3.
4.	5.	6.

Part 5 | Copy and work each problem.

P5			
a.	2 4 × 3	b. 5 1 × 3	c. 4 7 × 5

 **Lesson 84**

Part 6

$1.09 $2.93 $6.11 $2.06 $3.88

a. A boy buys juice, milk, and cereal.
About how much does he spend?

b. A boy buys juice, milk, and cereal.
Exactly how much does he spend?

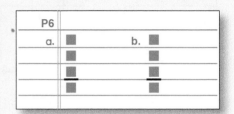

Part 7 Copy and work each problem.

a.
```
  430
 −191
```

b.
```
  500
 −330
```

c.
```
  660
 − 99
```

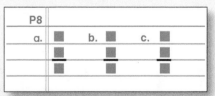

Part 8 For each problem, figure out the hours before 12. Add those hours to the hours after 12.

a. How many hours is it from 9 PM to 7 AM?

b. How many hours is it from 10 AM to 5 PM?

c. How many hours is it from 8 PM to 8 AM?

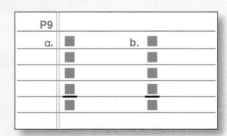

Part 9 Write the place-value addition fact for each number.

a. 3214   b. 4425

Lesson 84

Part 10 Work the column problem to answer each question.

> This graph shows how many trees were in each park.

P10		
a. ■		b. ■
	■		■
	■		■

Valley Park				
River Park				
Hill Park				
Central Park				

0 10 20 30 40 50
Trees

a. How many more trees were in Valley Park than in Central Park?

b. How many more trees were in River Park than in Hill Park?

Part 11 Copy each problem and write the answer.

P11		
a. $\dfrac{3}{4} - \dfrac{2}{4} = $ ___	b. $\dfrac{4}{7} + \dfrac{1}{7} = $ ___	c. $\dfrac{4}{7} - \dfrac{1}{7} = $ ___

Part 12 Write the cents amount with a dollar sign and a dot.

a. 40¢   b. 400¢   c. 4¢

P12		
a.	b.	c.

Connecting Math Concepts

Part 1

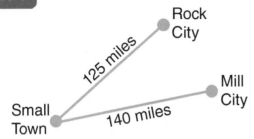

P1				
a.	b.	c. ■	d. ■	

a. John drove from Small Town to Rock City. How many miles did John travel?

b. Tom drove from Small Town to Mill City. How many miles did Tom travel?

c. How many more miles did Tom travel than John traveled?

d. How many miles did John and Tom travel altogether?

Part 2

a. There are 5 apples in every bag.
There are 20 bags.
How many apples are there?

b. There are 125 children at the park.
75 of the children are girls.
How many boys are at the park?

c. Each row has 4 desks.
There are 20 desks.
How many rows are there?

d. The store sold apples.
The store sold 125 red apples and 145 green apples.
How many apples did the store sell?

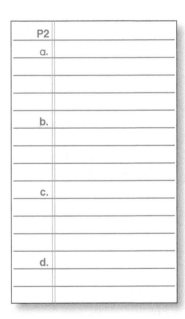

Part 3

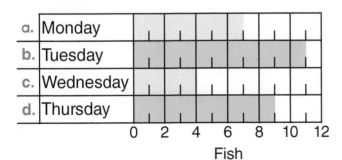

P3				
a.	b.	c.	d.	

Lesson 85

Part 4

	Monday	Tuesday	Wednesday
Henry	$165	$175	$145
Alice	$68	$86	$81
Mr. Black	$60	$80	$90

P4
a. ■ ■   b. ■ ■   c. ■ ■

a. How much more money did Henry earn on Tuesday than on Wednesday?

b. On Wednesday, how much more money did Mr. Black earn than he earned on Monday?

c. On Tuesday, how much more money did Henry earn than Mr. Black earned?

Independent Work

Part 5 Copy and work each problem.

P5
a.   1 3   b.   3 5   c.   3 0 1
   × 9   × 4   ×   4

Part 6 Write the answer to each item.

a. (hexagon)   b. (triangle)   c. (rectangle)   d. (tall rectangle)

e. (trapezoid)   f. (small square)   g. (pentagon)   h. (square)

1. Write the letter of each shape that is a pentagon.

2. Write the letter of each shape that is a hexagon.

3. Write the letter of each shape that is a triangle.

4. Write the letter of each shape that is a rectangle.

5. Write the letter of each shape that is a square.

6. Write the letter of each shape that is a quadrilateral.

P6
1.   2.   3.
4.   5.   6.

Lesson

Part 7 Copy each item and write the sign >, <, or =.

P7	
a. $\dfrac{4}{7}$ ■ $\dfrac{7}{7}$	b. $\dfrac{7}{3}$ ■ $\dfrac{7}{7}$ c. $\dfrac{3}{3}$ ■ $\dfrac{6}{6}$

Part 8 For each problem, figure out the hours before 12. Add those hours to the hours after 12.

a. How many hours is it from 10 AM to 5 PM?

b. How many hours is it from 11 AM to 4 PM?

c. How many hours is it from 8 PM to 7 AM?

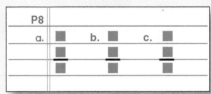

Part 9 For each item, write the column problem to find the missing number.

a. ■ − 300 = 300   b. ■ − 50 = 50   c. 200 + ■ = 275

P9			
a. ■	b. ■	c. ■	
■	■	■	
■	■	■	

Part 10 Write each number.

a. nine thousand nine hundred

b. nine thousand ninety

c. nine thousand nine

d. nine thousand

P10			
a.	b.	c.	d.

Lesson 85

Part 11 Work the column problem to answer each question.

This graph shows how many inches of rain fell each month.

January						
February						
March						
April						

0   2   4   6   8   10   12
Inches of Rain

P11			
a. ▪	b. ▪	c. ▪
▪	▪	▪
▪	▪	▪

a. How many more inches of rain were there in February than in January?

b. How many more inches of rain were there in April than in March?

c. How many more inches of rain were there in February than in April?

Part 12 Write the dollars and cents amount for each row.

a.

b.

P12			
a.		b.	c.

c.

Lesson

Part 1

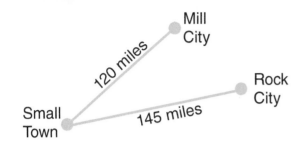

Mill City

120 miles

Rock City

Small Town

145 miles

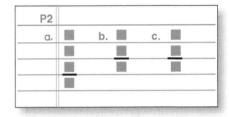

a. John drove from Small Town to Rock City. How many miles did John drive?

b. Tom drove from Small Town to Mill City. How many miles did Tom drive?

c. How many more miles did John drive than Tom drove?

d. How many miles did John and Tom drive altogether?

Part 2

	March	April	May
Judy	$425	$250	$190
Doug	$350	$325	$400
Pat	$150	$275	$275

a. How much money in all did the people earn in March?

b. In April, how much more money did Doug earn than Judy earned?

c. How much money did Doug earn in April and May?

Lesson

Part 3 For each item, make the number family and work the problem. Write the unit name in the answer.

a. Mario read 5 stories every day.
   He read 45 stories.
   How many days did he read stories?

b. Mary read 27 more pages than Don read.
   Mary read 134 pages.
   How many pages did Don read?

c. The duck flew 89 miles.
   The duck flew 112 fewer miles than the eagle flew.
   How far did the eagle fly?

d. There were 32 flowers in the garden.
   Each flower had 9 petals.
   How many petals were there in all?

P3	
a.	
b.	
c.	
d.	

Part 4 Write the fraction for each number line.

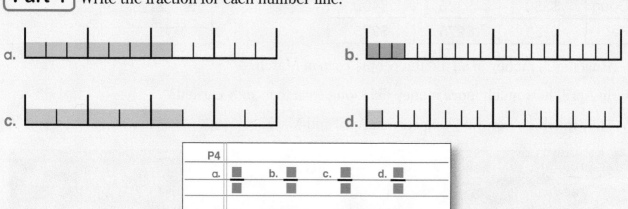

Part 5 Copy and work each problem.

P5			
a.	4 2 × 9	b. 2 7 × 5	c. 4 0 3 × 2

Part 6 | Write the answer to each question.

This graph shows how much fruit the store has.

Apples							
Bananas							
Oranges							
Pineapples							

0 10 20 30 40 50 60 70
Fruit

P6			
a.	■	b.	■
	■		■
	■		■

a. How many more apples than oranges does the store have?

b. How many more bananas than pineapples does the store have?

Part 7 | Write each number.

a. nine thousand nine

b. nine thousand

c. nine thousand nine hundred

d. nine thousand ninety

P7			
a.	b.	c.	d.

Part 8 | Copy each item and write the sign >, <, or =.

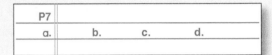

P8			
a. $\dfrac{7}{4}$ ■ $\dfrac{4}{7}$	b. $\dfrac{4}{5}$ ■ $\dfrac{5}{4}$	c. $\dfrac{8}{9}$ ■ $\dfrac{9}{8}$

Part 9 | For each item, write the column problem to find the missing number.

a. ■ − 30 = 30   b. 50 + ■ = 90   c. ■ − 30 = 40

P9			
a.	■	b. ■	c. ■
	■	■	■
	■	■	■

Lesson

		0	3	6	9	12	15
a.	Cars						
b.	Trucks						
c.	Motorcycles						
d.	Vans						

Vehicles

P1				
	a.	b.	c.	d.

Independent Work

Part 2 For each item, make the number family and work the problem.
Write the unit name in the answer.

a. At the start of the day, there were lots of cookies in a jar.
During the day, the children ate 59 of the cookies.
At the end of the day, there were 141 cookies in the jar.
How many cookies were in the jar to start with?

b. Kyle rode his bike for 2 hours every day.
He rode his bike for 8 days.
How long did Kyle ride his bike?

c. The children drew 69 pictures altogether.
Each child drew 3 pictures.
How many children were there?

d. The farmer had 144 rabbits.
The farmer bought some more rabbits.
The farmer ended up with 190 rabbits.
How many rabbits did the farmer buy?

P2	
a.	
b.	
c.	
d.	

Lesson 87

Part 3

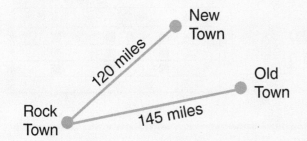

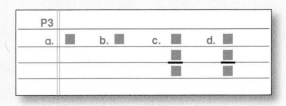

a. Ann drove from Rock Town to Old Town. How many miles did Ann drive?

b. Jane drove from Rock Town to New Town. How many miles did Jane drive?

c. How many more miles did Ann drive than Jane drove?

d. How many miles did Ann and Jane drive altogether?

Part 4 Copy and work each problem.

P4				
a.	37 × 2	b. 59 × 3	c. 30 × 9	d. 510 × 4

Part 5 Write the answer for each item.

1. Write the letter of each shape that is a quadrilateral.

2. Write the letter of each shape that is a rectangle.

3. Write the letter of each shape that is a square.

4. Write the letter of each shape that is a hexagon.

5. Write the letter of each shape that is a triangle.

6. Write the letter of each shape that is a pentagon.

P5			
1.		2.	3.
4.		5.	6.

Part 6 Copy and work each problem.

a.
$$\begin{array}{r} 3000 \\ -2500 \\ \hline \end{array}$$

b.
$$\begin{array}{r} 800 \\ -250 \\ \hline \end{array}$$

c.
$$\begin{array}{r} 600 \\ -350 \\ \hline \end{array}$$

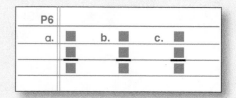

P6			
a.	■	b. ■	c. ■
	▬	▬	▬
	■	■	■

Part 7 Copy each item and write the sign >, <, or =.

P7			
a. $\dfrac{5}{3}$ ■ $\dfrac{2}{3}$	b. $\dfrac{7}{4}$ ■ $\dfrac{7}{10}$	c. $\dfrac{2}{9}$ ■ $\dfrac{9}{2}$	

Part 8 Copy and work each regular problem. Next to the regular problem, work an estimation problem with the numbers rounded to the nearest hundred.

P8			
a. $\begin{array}{r} 492 \\ +318 \\ \hline \end{array}$ $\begin{array}{r} ■ \\ +■ \\ \hline \end{array}$	b. $\begin{array}{r} 209 \\ +579 \\ \hline \end{array}$ $\begin{array}{r} ■ \\ +■ \\ \hline \end{array}$		

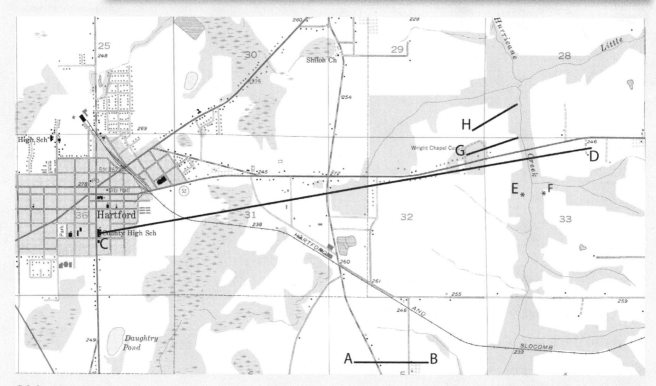

Lesson

a. The river was 4 times as deep as the pond.

b. The bird weighed 3 times as much as the frog.

c. Laura had 3 times as many books as Bonnie had.

d. Tom ran 4 times as many miles as Jake ran.

P1		
a.		b.
c.		d.

Part 2

		0 3 6 9 12 15
a.	Monday	
b.	Tuesday	
c.	Wednesday	
d.	Thursday	
e.	Friday	
f.	Saturday	

Fish

P2		
a.	b.	c.
d.	e.	f.

Independent Work

Part 3 For each item, make the number family and work the problem.
Write the unit name in the answer.

a. The goat weighs 69 pounds.
The pig weighs 82 pounds.
How much more does the pig weigh than the goat weighs?

b. There were 3 birds in every cage.
There were 43 cages.
How many birds were there in all?

c. Each box had 4 crayons in it.
There were 20 crayons in all.
How many boxes of crayons were there?

d. Alice ran 18 miles more than Tom ran.
Tom ran 42 miles.
How many miles did Alice run?

P3	
a.	
b.	
c.	
d.	

Lesson

Part 4

| $1.09 | $2.93 | $6.11 | $2.06 | $3.88 |

a. A man buys bananas, meat, and milk.
   About how much does he spend?

b. A boy buys bananas, meat, and milk.
   Exactly how much does he spend?

P4		
a. ▪	b. ▪	
▪	▪	
▪	▪	
▬	▬	

Part 5 Copy and work each problem.

P5			
a.	4 1 0	b. 3 2	c. 5 3
	× 9	× 5	× 4

Part 6

This table shows how much money each person earned during three months.

	May	June	July
Mario	$180	$125	$150
James	$170	$150	$240
Elisa	$130	$175	$270

P6		
a. ▪	b. ▪	c. ▪
▪	▪	▪
▬	▪	▬
	▬	
	▪	

a. How much more money did Elisa earn than Mario earned in June?

b. How much did Mario earn in the three months?

c. In May, how much more money did Mario earn than James?

Lesson

Part 7 Copy each problem and write the answer.

P7	
a. $\dfrac{7}{9} - \dfrac{3}{9} = $ ___	b. $\dfrac{3}{7} + \dfrac{2}{7} = $ ___ c. $\dfrac{4}{10} - \dfrac{3}{10} = $ ___

Part 8 Write the place-value addition fact for each number.

a. 8888   b. 4111

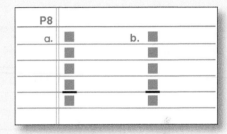

Part 9 For each problem, figure out the hours before 12. Add those hours to the hours after 12.

a. How many hours is it from 8 PM to 6 AM?

b. How many hours is it from 11 AM to 5 PM?

c. How many hours is it from 4 PM to 4 AM?

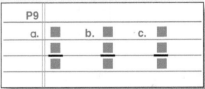

Part 10 Write the time for each clock.

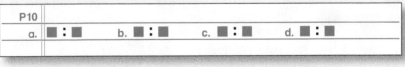

a.

b.

c.

d.

Lesson 89

Part 1

a. The redwood was 4 times as old as the elm.

b. The bedroom was 3 times as wide as the kitchen.

c. The yellow rock was 10 times as heavy as the white rock.

d. Mr. Jones earned 9 times as much money as Mr. Briggs.

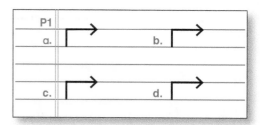

Part 2

These graphs show the age of children.

Graph 1

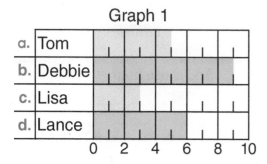

a.	Tom
b.	Debbie
c.	Lisa
d.	Lance

0 2 4 6 8 10

Graph 2

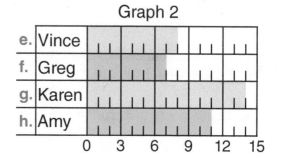

e.	Vince
f.	Greg
g.	Karen
h.	Amy

0 3 6 9 12 15

P2			
a.	b.	c.	d.
e.	f.	g.	h.

Connecting Math Concepts

Lesson 89

Part 3 For each item, make the number family and work the problem. Write the unit name in the answer.

a. Dan is 165 centimeters tall.
Dan is 19 centimeters shorter than Evan.
How many centimeters tall is Evan?

b. The cow weighed 932 pounds.
The horse weighed 714 pounds.
How much more did the cow weigh than the horse weighed?

c. Jerry spelled 32 words every minute.
He spelled words for 4 minutes.
How many words did he spell?

d. Every triangle has 3 angles.
There are 96 angles in all.
How many triangles are there?

P3	
a.	
b.	
c.	
d.	

Part 4

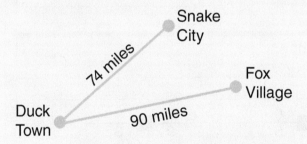

P4			
a. ■	b. ■	c. ■	d. ■
		■	■
		■	■

a. José drove from Snake City to Duck Town. How many miles did José drive?

b. Ann drove from Fox Village to Duck Town. How many miles did Ann drive?

c. How many more miles did Ann drive than José drove?

d. How many miles did José and Ann drive altogether?

Lesson 89

Part 5 Copy and work each problem.

P5			
a.	49 × 3	b. 501 × 9	c. 37 × 2

Part 6

a. Find the perimeter of this rectangle.

b. Find the area of this rectangle.

4 in.

3 in.

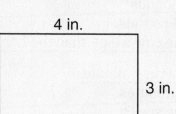

P6	
a.	■ + ■ = ■ ■ + ■ = ■ ■
b.	■ × ■ = ■

Part 7 Copy and work each problem.

a. 400
  − 70

b. 400
  − 50

c. 4000
  − 500

P7			
a. ■ ■ ■	b. ■ ■ ■	c. ■ ■ ■	

Part 8 Copy each item and write the sign >, <, or =.

P8			
a. $\frac{4}{4}$ ■ $\frac{3}{2}$	b. $\frac{4}{4}$ ■ $\frac{6}{6}$	c. $\frac{7}{9}$ ■ $\frac{3}{3}$	

Part 9 Write the fraction for each picture.

a.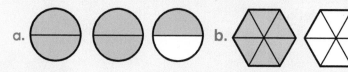

b.

c.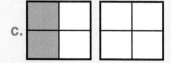

d.

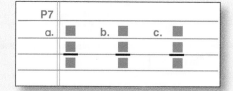

P9			
a. ■ ■	b. ■ ■	c. ■ ■	d. ■ ■

Connecting Math Concepts

Lesson 90

Part 1

a. There were 2 times as many blue cars as red cars.
   There were 46 blue cars.
   How many red cars were there?

b. There were 5 times as many apples as bananas.
   There were 30 bananas.
   How many apples were there?

P1	
a.	→
b.	→

Part 2

These graphs tell how many trees are in different parks.

Graph 1

Park A	
Park B	
Park C	
Park D	

0 3 6 9 12

Graph 2

Park E	
Park F	
Park G	
Park H	

0 2 4 6 8

P2				
a.	b.	c.	d.	
e.	f.	g.	h.	

Lesson

Part 3 For each item, make the number family and work the problem. Write the unit name in the answer.

a. There were 5 cats.
Each cat had 24 whiskers.
How many whiskers were there in all?

b. There were 72 children in the swimming pool.
Then some more children jumped into the pool.
Now there are 90 children in the pool.
How many children jumped into the pool?

c. Each team had 9 players.
There were 45 players in all.
How many teams were there?

P3	
a.	
b.	
c.	

Part 4 Write the answer to each question.

This graph shows how many miles a girl ran each month.

P4		
a.	b.	c. ■
		■
		■

April	🏃 🏃 🏃 🏃
May	🏃 🏃
June	🏃 🏃 🏃 🏃 🏃
July	🏃 🏃 🏃

🏃 = 10 miles

a. In which month did the girl run 20 miles?

b. How many miles did the girl run in May?

c. How many more miles did the girl run in April than in May?

Lesson 90

Part 5 Copy and work each problem.

P5				
a.	5 8 × 2	b. 1 4 0 × 2	c. 2 1 0 × 5	d. 3 4 × 3

Part 6 Copy each item and write the sign >, <, or =.

P6			
a.	$\dfrac{4}{3}$ ■ $\dfrac{8}{8}$	b. $\dfrac{5}{5}$ ■ $\dfrac{9}{10}$	c. $\dfrac{3}{3}$ ■ $\dfrac{8}{8}$

Part 7 For each item, work a column problem to find the missing number.

a. ■ − 40 = 40

b. 60 + ■ = 90

P7	
a.	b.

Part 8 Write each number.

a. six thousand six hundred

b. six thousand

c. six thousand sixty

d. six thousand six

P8			
a.	b.	c.	d.

Part 9 For each problem, figure out the hours before 12. Add those hours to the hours after 12.

a. How many hours is it from 10 PM to 7 AM?

b. How many hours is it from 8 AM to 3 PM?

c. How many hours is it from 5 PM to 2 AM?

P9			
a. ▪	b. ▪	c. ▪	
▪	▪	▪	
▪	▪	▪	

Part 10 Work the column problem for each item.

a. What is 7 dollars minus $4.30?

b. What is 5 dollars minus $.80?

P10		
a. ▪	b. ▪	
▪	▪	
▪	▪	

Lesson

a. The yellow snake was 9 times as long as the red snake.
The red snake was 10 inches long.
How long was the yellow snake?

b. There were 6 times as many spoons as forks.
There were 66 spoons.
How many forks were there?

c. The pine tree was 5 times as tall as the maple tree.
The pine tree was 105 feet tall.
How many feet tall was the maple tree?

d. The cat weighed 15 pounds.
The dog weighed 3 times as much as the cat.
How much did the dog weigh?

P1	→
a.	
b.	
c.	
d.	

Part 2

a. 20 × 7   b. 30 × 4   c. 90 × 3

d. 50 × 9   e. 30 × 5   f. 20 × 6

P2		
a.	b.	c.
d.	e.	f.

Lesson

Part 3 | For each item, make the number family and work the problem. Write the unit name in the answer.

P3	
a.	
b.	
c.	
d.	

a. The maple tree was 34 feet tall.
The oak tree was 40 feet taller than the maple tree.
How tall was the oak tree?

b. Each car has 4 seats.
There are 6 cars.
How many seats are there?

c. The worm was 14 centimeters long.
The snake was 60 centimeters long.
How much shorter was the worm than the snake?

d. Jill reads 4 pages every day.
How many days will it take Jill to read 24 pages?

Part 4 | Work the column problem for each item.

a. How much is 24 dollars minus $9.30?

b. How much is 5 dollars minus $3.10?

P4			
a.	■	b.	■
	■		■
	■		■

Part 5 | Answer the questions about the graph.

This graph shows how many days it rained each month.

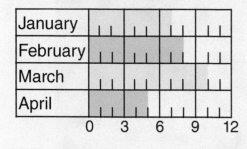

P5			
a.		b.	c. ■
			■
			■

a. In which month did it rain 5 days?

b. How many days did it rain in February?

c. How many more days did it rain in March than in April?

Part 6 Write the missing number in each row.

    a. 60 90 ■ 150

    b. 40 80 ■ 160

    c. 50 100 ■ 200

P6		
a.	b.	c.

Part 7 Copy and work each problem.

a.   3 0 1   b. 2 5   c. 9 3   d. 2 0

   ×   2   × 3   × 4   × 4

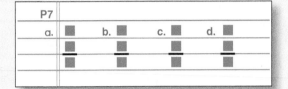

Part 8 Work column problems to answer questions about the table.

This table shows how much money each person earned during three months.

	October	November	December
Mark	$50	$100	$150
Lee	$75	$150	$150
Alicia	$100	$175	$150

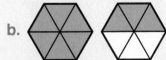

    a. How much did the three people earn in December?

    b. How much did Mark earn in the three months?

    c. In November, how much more money did Alicia earn than Mark earned?

Part 9 Write the fraction for each picture.

a.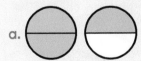

b.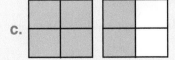

c.

d.

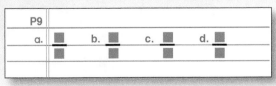

Lesson 91

Part 10

Copy and work each regular problem. Next to the regular problem, work an estimation problem with the numbers rounded to the nearest hundred.

P10						
a.	398	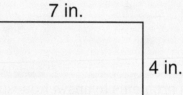		b.	274	
	+413	+			+536	+

Part 11

a. Find the perimeter of this rectangle.

b. Find the area of this rectangle.

7 in.

4 in.

P11	
a.	■ + ■ = ■
	■ + ■ = ■
	■
b.	■ × ■ = ■

Lesson 92

Part 1

a. The red coat costs 3 times as much as the brown coat.
The red coat costs 120 dollars.
How many dollars does the brown coat cost?

b. The rectangle is 4 times as long as the square.
The square is 20 centimeters long.
How many centimeters long is the rectangle?

c. The cow weighed 639 pounds.
The cow was 3 times as heavy as the goat.
How many pounds did the goat weigh?

d. The fir tree is 30 feet tall.
The pine tree is 4 times as tall as the fir tree.
How many feet tall is the pine tree?

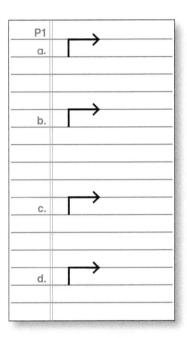

Part 2

a. 50 × 3   b. 40 × 9   c. 90 × 2

d. 50 × 7   e. 30 × 4   f. 50 × 5

P2		
a.	b.	c.
d.	e.	f.

Part 3

a. 6 hours and 16 minutes   d. 11 hours

b. 49 minutes   e. 9 hours and 2 minutes

c. 7 minutes   f. 4 minutes

P3	
a.	d.
b.	e.
c.	f.

Part 4

P4			
a. 5⟌350	b. 2⟌144	c. 2⟌804	

Lesson 92

Independent Work

Part 5 | For each item, make the number family and work the problem.

a. There were 9 players on each team.
There were 99 players altogether.
How many teams were there?

b. The cow weighs 642 pounds.
The cow is 130 pounds heavier than the horse.
How much does the horse weigh?

c. Mrs. Vernon had 4 boxes of cookies.
Each box had 14 cookies.
How many cookies did she have in all?

P5	
a.	
b.	
c.	

Part 6 | Work the column problem for each item.

a. How much is 8 dollars minus $2.50?

b. How much is 47 dollars minus $12.30?

c. How much is 7 dollars minus $3.50?

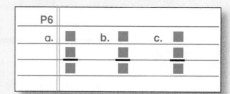

Part 7 | Copy each problem and write the answer.

P7			
a. $\dfrac{4}{5} - \dfrac{2}{5} =$ __	b. $\dfrac{5}{8} + \dfrac{2}{8} =$ __	c. $\dfrac{2}{5} - \dfrac{1}{5} =$ __	

Part 8 | Copy and work each problem.

a. $\begin{array}{r} 1\,1 \\ \times\ \ 9 \\ \hline \end{array}$ b. $\begin{array}{r} 5\,1\,0 \\ \times\ \ \ \ 3 \\ \hline \end{array}$ c. $\begin{array}{r} 1\,6 \\ \times\ \ 4 \\ \hline \end{array}$ d. $\begin{array}{r} 2\,5 \\ \times\ \ 7 \\ \hline \end{array}$

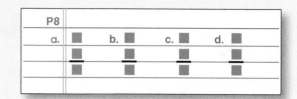

Connecting Math Concepts

Lesson 92

Part 9 | Write the missing number for each item.

a. 180 270 ■ 450

b. 250 300 ■ 400

c. 120 140 ■ 180

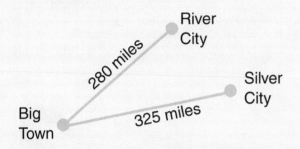

Part 10 | Answer the questions about the map.

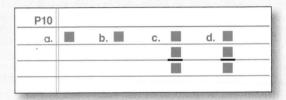

a. Alice drove from Big Town to River City. How many miles did Alice drive?

b. Sara drove from Big Town to Silver City. How many miles did Sara drive?

c. How many more miles did Sara drive than Alice drove?

d. How many miles did Alice and Sara drive altogether?

Part 11 | Answer the questions about the graph.

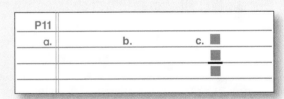

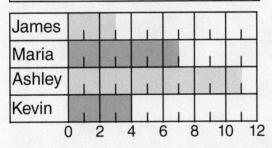

a. How many miles did James run?

b. How many miles did Ashley run?

c. How many more miles did Ashley run than Maria ran?

Lesson 92

Part 12 For each rectangle, write a times problem and figure out the number of squares.

a.

b.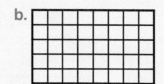

P12	
a.	b.

Part 13 Write the cents amount with a dollar sign and a dot.

a. 300¢　　b. 9¢　　c. 45¢　　d. 620¢

P13			
a.	b.	c.	d.

Lesson 93

Part 1

a. There are 3 times as many children in School A as in School B.
There are 360 children in School A.
How many children are in School B?

b. The pine tree was 4 times as tall as the maple tree.
The maple tree was 23 feet tall.
How many feet tall was the pine tree?

c. There were 3 times as many spoons as forks.
There were 15 spoons.
How many forks were there?

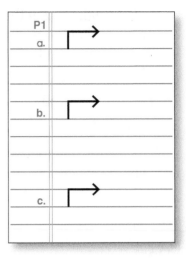

Part 2

a. 50 × 7   b. 90 × 4   c. 40 × 6

d. 90 × 5   e. 30 × 3   f. 20 × 6

Part 3

a. 6 hours + 2 hours and 8 minutes

b. 3 hours and 9 minutes + 14 minutes

c. 4 hours and 19 minutes − 7 minutes

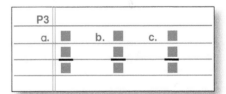

Independent Work

Part 4 Copy and work each problem.

a. 4)248   b. 3)150   c. 3)963   d. 2)180

Lesson 93

Part 5 For each item, make the number family and work the problem. Write the unit name in the answer.

a. At the beginning of the month, the cow weighed 426 pounds.
At the end of the month, the cow weighed 470 pounds.
How many pounds did the cow gain during the month?

b. The boat was 38 feet long.
The truck was 14 feet shorter than the boat.
How long was the truck?

c. There are 22 students in each class.
There are 4 classes.
How many students are there?

P5	
a.	
b.	
c.	

Part 6 Write the fraction for each number line.

a.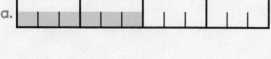

b.

c.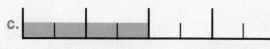

d.

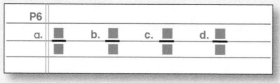

Part 7 Answer the questions about the graph.

This graph shows how many apples Gina ate each month.

P7			
a.		b.	c. ■
			■
			■

April						
May						
June						
July						
	0	3	6	9	12	15

a. How many apples did Gina eat in May?

b. How many apples did Gina eat in April?

c. How many more apples did Gina eat in June than in July?

Lesson

Part 8 Copy and work each problem.

a.
```
  400
-  50
```

b.
```
  4000
-  500
```

c.
```
  425
-  77
```

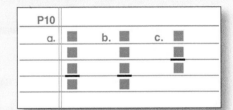

Part 9 Write the missing number for each item.

a. 350 400 ■ 500

b. 120 160 ■ 240

c. 120 150 ■ 210

Part 10 Work column problems to answer questions about the table.

This table shows how much money each person earned during three months.

	May	June	July
James	$150	$125	$150
Alex	$140	$125	$240
Rose	$160	$125	$250

a. How much did the three people earn in May?

b. How much did James earn in the three months?

c. In July, how much more money did Rose earn than James earned?

Part 11 Copy and work each problem.

a.
```
  25
×  7
```

b.
```
  58
×  2
```

c.
```
  36
×  4
```

d.
```
  200
×   3
```

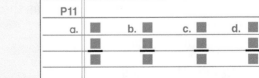

Lesson 94

Part 1

a. 7×50 b. 4×30 c. 6×20

d. 3×90 e. 50×7 f. 20×4

P1			
a.		b.	c.
d.		e.	f.

Part 2

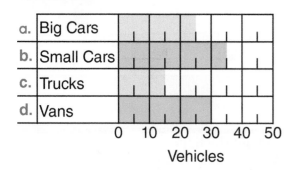

a. Big Cars
b. Small Cars
c. Trucks
d. Vans

0 10 20 30 40 50

Vehicles

P2			
a.	b.	c.	d.

Part 3

a. 28 minutes + 3 hours and 7 minutes

b. 6 hours and 58 minutes – 5 hours

c. 2 hours and 30 minutes + 1 hour and 21 minutes

d. 3 hours and 7 minutes – 2 hours and 2 minutes

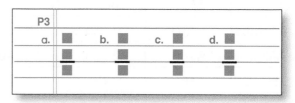

Independent Work

Part 4 For each item, make the number family and work the problem.
Write the unit name in the answer.

a. The cow weighed 3 times as much as the pig.
The pig weighed 231 pounds.
How much did the cow weigh?

b. Alice found 60 seashells.
Alice found 2 times as many seashells as Gary found.
How many seashells did Gary find?

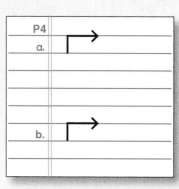

Lesson 94

Part 5

Write the missing number in each equation.

a. 2 × ■ = 16 b. 5 × ■ = 35 c. 4 × ■ = 24

P5			
	a.	b.	c.

Part 6

Copy and work each problem.

P6				
	a. $4\overline{)204}$	b. $5\overline{)350}$	c. $3\overline{)276}$	d. $2\overline{)868}$

Part 7

Copy and work each regular problem. Next to the regular problem, work an estimation problem with the numbers rounded to the nearest hundred.

P7				
a.	229 +185	■ +■	b. 388 +488	■ +■

Part 8

a. Find the perimeter of this rectangle.

b. Find the area of this rectangle.

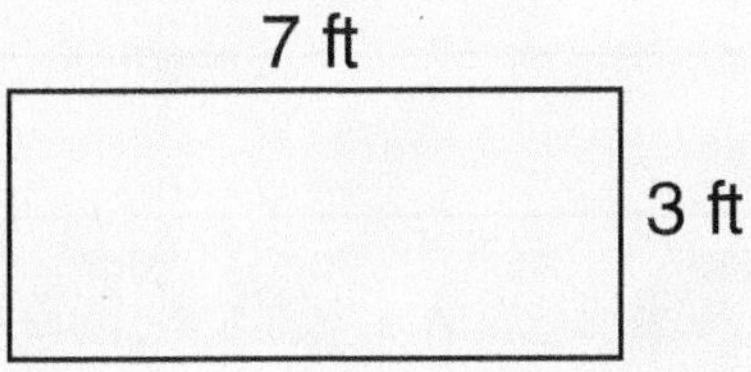

P8	
a.	■ + ■ = ■
	■ + ■ = ■/■
b.	■ × ■ = ■

Lesson 94

Part 9
For each item, make the number family and work the problem. Write the unit name in the answer.

a. Each basket weighed 2 pounds.
All the baskets together weighed 24 pounds.
How many baskets were there?

b. There were two coins in every box.
There were 56 boxes.
How many coins were there?

c. There were 145 passengers on the train.
Some more passengers got on the train.
The train ended up with 195 passengers.
How many more passengers got on the train?

P9	
a.	
b.	
c.	

Part 10
Copy each item and write the sign >, <, or =.

P10			
a. $\frac{5}{8}$ ■ $\frac{4}{4}$	b. $\frac{6}{6}$ ■ $\frac{3}{3}$	c. $\frac{3}{2}$ ■ $\frac{6}{6}$	

Part 11
Write the answer to each question.

P11		
a.	b.	c.

This graph shows how many houses are on each street.

James Street	🏠
Adams Street	🏠 🏠 🏠 🏠 🏠
High Street	🏠 🏠
River Street	🏠 🏠 🏠

🏠 = 5 houses

a. Which street has 5 houses?

b. How many more houses are on River Street than on James Street?

c. Which street has more than 15 houses?

Lesson 95

Part 1

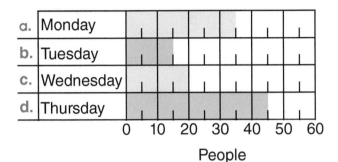

a.	Monday		
b.	Tuesday		
c.	Wednesday		
d.	Thursday		

0 10 20 30 40 50 60

People

P1	
a. b. c. d.	

Part 2

a. John leaves for work at 7:13.
He drives for 1 hour and 6 minutes.
When does he arrive at work?

b. The movie started at 4:36.
The movie was 2 hours long.
When did the movie end?

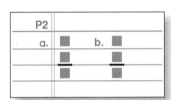

P2	
a. ■ b. ■	
■ ■	
▬ ▬	

Part 3

a. 4 × 30   b. 9 × 50   c. 30 × 5

d. 6 × 20   e. 7 × 50   f. 40 × 4

P3	
a. b. c.	
d. e. f.	

Independent Work

Part 4 For each item, make the number family and work the problem.
Write the unit name in the answer.

a. The shirt cost 15 dollars.
The coat cost 3 times as much as the shirt.
How much did the coat cost?

b. The redwood tree is 5 times as tall as the cherry tree.
The redwood tree is 150 feet tall.
How tall is the cherry tree?

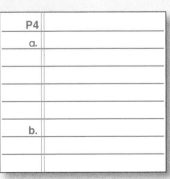

P4	
a.	
b.	

Lesson 95

Part 5 | Copy and work the problems.

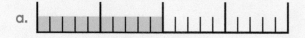

Part 6 | Write the fraction for each number line.

a.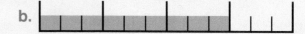

b.

c.

d.

Part 7 | For each item, make the number family and work the problem.
Write the unit name in the answer.

a. Mike spent 84 cents to buy pencils.
   Each pencil cost 4 cents.
   How many pencils did Mike buy?

b. The turtle was 113 years old.
   The lizard was 56 years old.
   How much older was the turtle than the lizard?

c. A school had 414 students at the beginning of the
   school year.
   29 students left the school.
   How many students were still in the school?

d. Libby read for 35 minutes each day.
   Libby read for 4 days.
   How many minutes did Libby read in all?

Part 8 For each problem, figure out the hours before 12. Add those hours to the hours after 12.

a. How many hours is it from 8 PM to 6 AM?

b. How many hours is it from 7 AM to 2 PM?

P8	
a. ■	b. ■
■	■
■	■

Part 9 Write the place-value addition fact for each number.

a. 1412   b. 2852

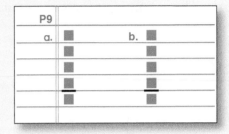

Part 10 Write the missing number in each equation.

a. $5 \times \blacksquare = 40$ b. $9 \times \blacksquare = 36$ c. $4 \times \blacksquare = 28$

P10		
a.	b.	c.

Part 11 Work the problem.

a. The man had 6 dollars.
He spent $3.40.
How much money does he have left?

P11	
a. ■	
■	
■	

Part 12 For each rectangle, write a times problem and figure out the number of squares.

a.   b.

P12	
a.	b.

Lesson 96

Part 1

a. The snake was how many times as old as the rabbit?

b. The elephant was how many times as heavy as the fox?

c. The number of adults was how many times the number of children?

d. The mountain was how many times as tall as the hill?

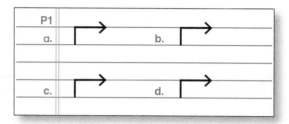

Part 2

a. 4 × 50

b. 90 × 2

c. 6 × 40

d. 4 × 30

e. 40 × 4

f. 7 × 20

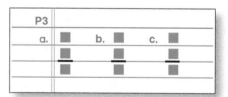

Part 3

a. The alarm started ringing at 8:21.
The alarm rang for 1 hour and 8 minutes.
When did the alarm stop ringing?

b. The plane to Chicago left at 7 o'clock.
The trip took 3 hours and 51 minutes.
When did the plane arrive in Chicago?

c. Juanita started reading the book at 3 o'clock.
She read for 25 minutes.
When did she stop reading?

Connecting Math Concepts

Lesson 96

Part 4

These graphs tell how many benches are in different parks.

Graph 1

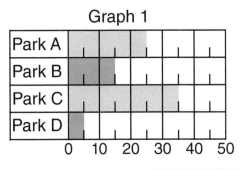

Graph 2

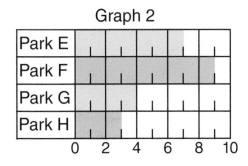

P4				
a.		b.	c.	d.
e.		f.	g.	h.

Independent Work

Part 5 For each item, make the number family and work the problem. Write the unit name in the answer.

a. Kevin ran 4 times as many miles as James ran last month.
James ran 35 miles.
How many miles did Kevin run?

b. There are 5 times as many roses as daisies in the yard.
There are 150 roses in the yard.
How many daisies are there in the yard?

c. 25 boys in our school play football.
3 times as many boys play baseball as play football.
How many boys in our school play baseball?

P5	
a.	
b.	
c.	

Part 6 Copy and work each problem.

P6			
a. $3\overline{)153}$	b. $9\overline{)180}$	c. $2\overline{)862}$	

Lesson 96

Part 7 | Write the place-value addition fact for each number.

a. 7421

b. 1272

P7			
a.	■	b.	■
	■		■
	■		■
	■		■

Part 8 | Work each problem. Write the unit name.

a. The girl had 17 fewer marbles than the boy had.
The girl had 113 marbles.
How many marbles did the boy have?

b. In April, Margie read 170 pages.
In May, she read 329 pages.
How many more pages did Margie read in May than in April?

P8	
a.	
b.	

Part 9 | Round each number to the nearest ten.

a. 72   b. 37   c. 58

d. 52   e. 14

P9					
a.	b.	c.	d.	e.

Part 10 | Copy and work each problem.

a. 4 7
   × 3

b. 5 9
   × 3

c. 4 1 0
   ×   5

P10					
a.	■	b. ■	c.	■
	■		■		■
	■		■		■

Lesson 96

Part 11 Write the time for each clock.

P11				
a. ■ : ■	b. ■ : ■	c. ■ : ■	d. ■ : ■

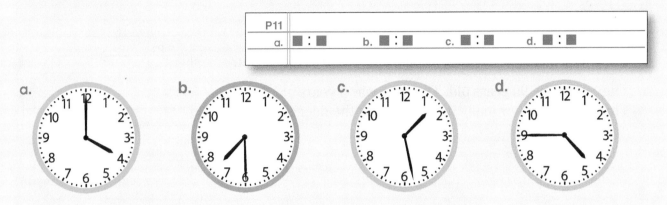

a.   b.   c.   d.

Part 12 Work each problem. Write the unit name.

a. There are 25 students in each classroom.
There are 7 classrooms.
How many students are there?

b. The store sells 4 shirts every day.
How many days will it take the store to sell 120 shirts?

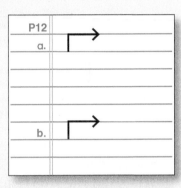

Lesson

a. The red hat costs 9 dollars. The blue hat costs 45 dollars.
   The blue hat costs how many times as much as the red hat?

b. The turtle was 90 years old. The deer was 3 years old.
   The turtle was how many times as old as the deer?

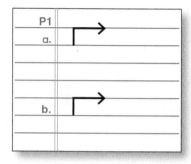

Independent Work

Part 2 Work a column problem for each item. Remember to write the dots.

a. The train to Boston left at 8 o'clock.
   The trip took 2 hours and 15 minutes.
   When did the train arrive in Boston?

b. Edna started reading the book at 7:15.
   She read for 30 minutes.
   When did she stop reading?

Part 3 Write the answer to each problem.

a. 5 × 70   b. 90 × 5   c. 2 × 60   d. 50 × 3

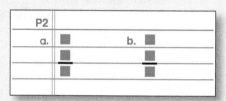

Part 4 For each item, make the number family and work the problem.
   Write the unit name in the answer.

a. The goat weighed 3 times as much as the dog.
   The goat weighed 150 pounds.
   How much did the dog weigh?

b. The redwood tree was 4 times as tall as the apple tree.
   The redwood tree was 128 feet tall.
   How tall was the apple tree?

Part 5 | Copy and work each problem.

P5			
a. $4\overline{)240}$	b. $3\overline{)120}$	c. $2\overline{)862}$	

Part 6 | Work the column problem for each item.

a. What is 9 dollars minus $3.50?

b. What is 24 dollars minus $10.30?

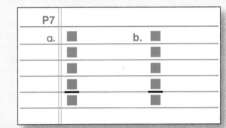

Part 7 | Write the place-value addition fact for each number.

a. 1524   b. 9916

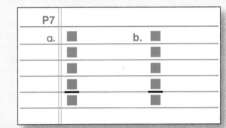

Part 8 | Copy and work each problem.

a. $\begin{array}{r} 950 \\ -325 \\ \hline \end{array}$ b. $\begin{array}{r} 470 \\ -190 \\ \hline \end{array}$ c. $\begin{array}{r} 5250 \\ +950 \\ \hline \end{array}$

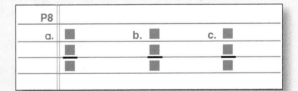

Part 9 | For each item, make the number family and work the problem. Write the unit name in the answer.

a. There were 5 girls on each team.
There were 150 girls altogether.
How many teams were there?

b. Each bag had 9 apples.
There were 25 bags of apples.
How many apples were there?

Lesson 97

Part 10 Write the answer to each question.

This graph shows how many pages of a book each child read.

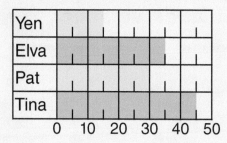

P10			
a.		b.	c. ■
			■
			▬

a. Who read the most pages?

b. How many pages did Tina read?

c. How many more pages did Elva read than Pat read?

Part 11

7 in.

5 in.

a. Find the perimeter of this rectangle.

b. Find the area of this rectangle.

P11	
a.	■ + ■ = ■
	■ + ■ = ■
	■
b.	■ × ■ = ■

Part 12 Write the numbers for counting by 9s to 90.

P12	
a.	9 ■ ■ ■ ■ ■ ■ ■ 90

Connecting Math Concepts

Lesson

a. The dog slept for 4 hours. The bear slept for 20 hours.
   The bear slept how many times as long as the dog slept?

b. Rita read 36 pages. Joe read 9 pages.
   Rita read how many times as many pages as Joe read?

c. There were 30 children on the swings.
   There were 3 adults on the swings.
   There were how many times as many children as adults on
   the swings?

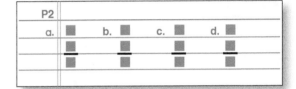

Part 2

a. 6 2
   × 3

b. 7 1
   × 5

c. 8 9
   × 2

d. 3 8
   × 4

Independent Work

Part 3 For each item, make the number family and work the problem.
Write the unit name in the answer.

a. The ship started out with 294 passengers on it.
   Some passengers got off the ship.
   The ship ended up with 119 passengers.
   How many passengers got off the ship?

b. On Thursday, the store sold 86 more cups than plates.
   The store sold 123 plates.
   How many cups did the store sell?

c. Tomás paints 4 pictures every week.
   How many weeks will it take him to paint 120 pictures?

Lesson 98

Part 4 — Copy each item and write the sign >, <, or =.

P4			
a. $\dfrac{4}{4}$ ■ $\dfrac{5}{8}$	b. $\dfrac{4}{3}$ ■ $\dfrac{5}{5}$	c. $\dfrac{6}{6}$ ■ $\dfrac{8}{8}$	

Part 5 — Work a column problem for each item. Remember to write the dots.

a. The plane to New York left at 9 o'clock.
The trip took 1 hour and 5 minutes.
When did the plane arrive in New York?

b. Kevin started reading the book at 4 o'clock.
He read for 45 minutes.
When did he stop reading?

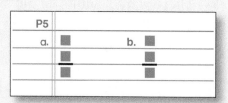

Part 6 — Write the cents amount with a dollar sign and a dot.

a. 40¢   b. 4¢   c. 400¢

P6			
a.	b.	c.	

Part 7 — Write the missing number for each item.

a. 120   140   ■   180

b. 120   160   ■   240

c. 120   150   ■   210

P7			
a.	b.	c.	

Lesson 98

Part 8 | Write the answer to each question.

This graph shows how many inches of
snow fell during four months.

December	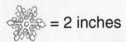
January	
February	
March	

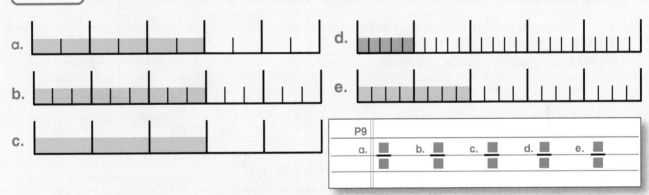

= 2 inches

a. How many inches of snow fell
   in December?

b. In which month did 4 inches of snow fall?

c. How many more inches of snow fell
   in February than in March?

P8			
a.		b.	c.

Part 9 | Write the fraction for each number line.

a.

b.

c.

d.

e.

P9					
a. ■	b. ■	c. ■	d. ■	e. ■
■	■	■	■	■

Part 10 | Copy and work each problem.

a. $\begin{array}{r} 37 \\ \times\ 4 \\ \hline \end{array}$ b. $\begin{array}{r} 640 \\ \times\ \ \ 2 \\ \hline \end{array}$ c. $\begin{array}{r} 19 \\ \times\ 3 \\ \hline \end{array}$

P10			
a. ■	b. ■	c. ■
■	■	■
■	■	■

Part 11 | Write the numbers for counting by 9s to 90.

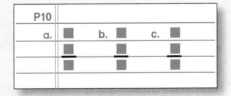

P11	
a.	9 ■ ■ ■ ■ ■ ■ ■ ■ 90

Lesson

a. 18
   × 3

b. 49
   × 2

c. 31
   × 8

d. 22
   × 5

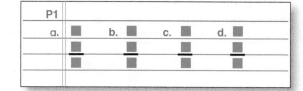

Part 2

a. The motel had 44 beds. The house had 4 beds.
   The motel had how many times as many beds as the house?

b. Maple Road was 10 times as long as Hill Road.
   Hill Road was 5 miles long.
   How many miles long was Maple Road?

c. Donna ate 9 peanuts. Greg ate 36 peanuts.
   Greg ate how many times as many peanuts as Donna ate?

d. The pig weighed 3 times as much as the sheep.
   The sheep weighed 60 pounds.
   How many pounds did the pig weigh?

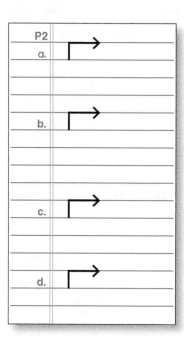

Independent Work

Part 3 For each item, make the number family and work the problem.
Write the unit name in the answer.

a. Each basket weighed 2 pounds.
   All the baskets together weighed 144 pounds.
   How many baskets were there?

b. There are 5 boxes of pencils.
   Each box has 25 pencils.
   How many pencils are there?

c. There were 9 cows in each barn.
   There were 45 cows in all.
   How many barns were there?

Lesson 99

Part 4 Copy and work each problem.

a.
$$
\begin{array}{r}
846 \\
-\ 90 \\
\hline
\end{array}
$$

b.
$$
\begin{array}{r}
1842 \\
+\ \ \ 58 \\
\hline
\end{array}
$$

c.
$$
\begin{array}{r}
430 \\
-295 \\
\hline
\end{array}
$$

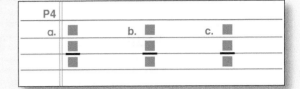

Part 5 Write the answer to each problem.

a. 20×6　b. 9×40　c. 5×80　d. 40×3

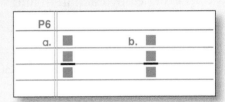

Part 6 Write a column problem for each item. Remember to write the dots.

a. The plane to New York left at 5 o'clock.
The trip took 4 hours and 9 minutes.
When did the plane arrive in New York?

b. It started raining at 4:15.
It rained for 30 minutes.
At what time did it stop raining?

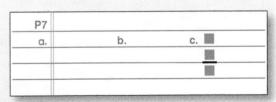

Part 7 Answer the questions about the graph.

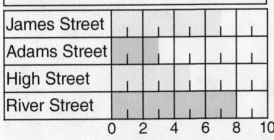

This graph shows how many houses are on each street.

James Street
Adams Street
High Street
River Street
0　2　4　6　8　10

a. Which street has 5 houses?

b. How many houses are on Adams Street?

c. How many more houses are on High Street than on Adams Street?

Part 8 Copy and work each problem.

P8　a. $9\overline{)360}$　b. $4\overline{)164}$　c. $2\overline{)608}$

Lesson 99

Part 9 | Copy each fraction and write the whole number the fraction equals.

P9	
a. $\dfrac{12}{2}$ = b. $\dfrac{15}{5}$ = c. $\dfrac{6}{3}$ =	

Part 10 | Work the column problem for each item.

a. What is 8 dollars minus $5.50?

b. What is 7 dollars minus $3.40?

P10		
a. ■	b. ■	
■	■	
■	■	

Part 11 | Copy each item and write the sign >, <, or =.

P11	
a. 7 × 5 ■ 30 + 10 b. 40 + 5 ■ 9 × 5	

Lesson

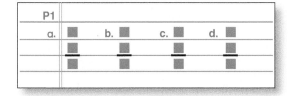

Part 1

a.   2 3
   × 5
‾‾‾‾‾

b.   4 7
   × 3
‾‾‾‾‾

c.   6 1
   × 4
‾‾‾‾‾

d.   5 8
   × 3
‾‾‾‾‾

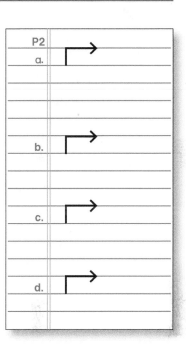

Part 2

a. Jimmy had 4 times as many dollars as Ellen had.
Ellen had 14 dollars.
How many dollars did Jimmy have?

b. The building had 36 windows and 9 doors.
The building had how many times as many windows as doors?

c. The cat weighed 20 pounds. The kitten weighed 5 pounds.
The cat was how many times as heavy as the kitten?

d. The brown house was 4 times as old as the green house.
The green house was 12 years old.
How many years old was the brown house?

Part 3

a. $4 \times 3 \times 5 =$ b. $7 \times 5 \times 3 =$

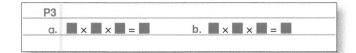

Lesson

Part 4 | For each item, make the number family and work the problem.
Write the unit name in the answer.

a. Mary started out with 56 cards.
She gave away some cards and ended up with 11 cards.
How many did she give away?

b. The truck started out with lots of packages.
It picked up 302 packages.
The truck ended up with 607 packages.
How many packages did it start out with?

c. The store had 58 candles.
A person bought a lot of candles.
The store ended up with 4 candles.
How many candles did the person buy?

d. Every day Jean reads 9 pages of a book.
How many days will it take Jean to read 180 pages?

P4	
a.	
b.	
c.	
d.	

Part 5 | Copy each fraction and write the whole number the fraction equals.

P5			
a. $\dfrac{6}{2} =$	b. $\dfrac{8}{4} =$	c. $\dfrac{12}{3} =$	

Part 6 | Write the missing number for each item.

a. 180 270 ■ 450

b. 180 210 ■ 270

c. 180 200 ■ 240

P6		
a.	b.	c.

Lesson 100

Part 7 Copy and work each problem.

P7			
a. $5\overline{)350}$	b. $4\overline{)368}$	c. $2\overline{)162}$	

Part 8 Copy each item and write the sign >, <, or =.

P8			
a. $\dfrac{8}{8}$ ▪ $\dfrac{9}{10}$	b. $\dfrac{6}{6}$ ▪ $\dfrac{4}{3}$	c. $\dfrac{4}{4}$ ▪ $\dfrac{8}{8}$	

Part 9

a. Find the perimeter of this rectangle.

b. Find the area of this rectangle.

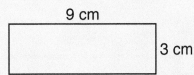

9 cm

3 cm

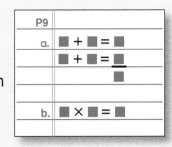

P9	
a.	▪ + ▪ = ▪
	▪ + ▪ = ▪
	▪
b.	▪ × ▪ = ▪

Part 10 Write the answer to each place-value problem.

a. 5000 + 400 + 10 + 4   b. 2000 + 600 + 40 + 1

c. 7000 + 200 + 10 + 8   d. 3000 + 900 + 70 + 1

P10			
a.	b.	c.	d.

Lesson 100

Part 11 Copy and work each problem.

a. 37
   × 4

b. 80
   × 3

c. 17
   × 5

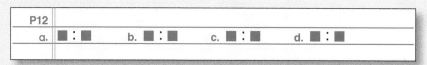

P11			
a. ■	b. ■	c. ■	
■	■	■	
■	■	■	

Part 12 Write the time for each clock.

P12			
a. ■ : ■	b. ■ : ■	c. ■ : ■	d. ■ : ■

a.

b.

c.

d.

Lesson 101

Part 1

a. It took 2 hours and 15 minutes to fix the car.
   It took 30 minutes longer to fix the truck than to fix the car.
   How long did it take to fix the truck?

b. On Monday, the woman painted the house for 2 hours and 30 minutes.
   On Tuesday, the woman painted for 1 hour and 5 minutes.
   How much longer did she paint on Monday than on Tuesday?

P1	
a.	⟶
b.	⟶

Part 2

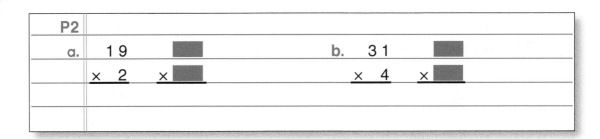

P2	
a.	1 9 ▮ b. 3 1 ▮
	× 2 × ▮ × 4 × ▮

Independent Work

Part 3 For each item, make the number family and work the problem.

a. Jerry did 4 times as many math problems as Gary did.
   Gary did 22 math problems.
   How many math problems did Jerry do?

b. The cat weighed 12 pounds.
   The dog weighed 5 times as much as the cat.
   How much did the dog weigh?

c. Elisa ran 15 miles last week.
   Dylan ran 3 miles last week.
   Elisa ran how many times as many miles as Dylan ran?

P3	
a.	
b.	
c.	

Lesson 101

Part 4 | Copy and work each problem.

    a. 3 4   b. 4 3   c. 5 6

        × 5   × 9   × 2

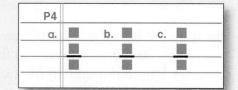

Part 5 | Copy and work each problem.

a. $9 \times 2 \times 5 =$ b. $3 \times 4 \times 2 =$

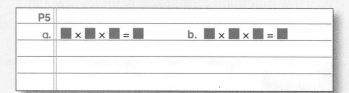

Part 6 | Work the column problem for each item.

a. What is 9 dollars minus $3.20?

b. What is 5 dollars minus $1.50?

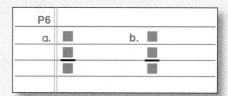

Part 7 | Copy and work each problem.

a. $3\overline{)156}$ b. $2\overline{)128}$ c. $4\overline{)84}$

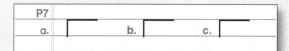

Part 8 | Copy each problem and write the answer.

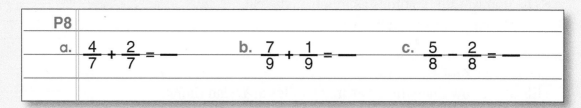

a. $\dfrac{4}{7} + \dfrac{2}{7} = \underline{}$ b. $\dfrac{7}{9} + \dfrac{1}{9} = \underline{}$ c. $\dfrac{5}{8} - \dfrac{2}{8} = \underline{}$

          Connecting Math Concepts

Lesson 101

Part 9 Write the missing number in each row.

a. 60 90 ▮ 150

b. 40 80 ▮ 160

c. 50 100 ▮ 200

P9		
a.	b.	c.

Part 10 For each item, make the number family and work the problem. Write the unit name in the answer.

a. Gary had 145 baseball cards.
He bought some more baseball cards.
He ended up with 235 baseball cards.
How many baseball cards did Gary buy?

b. Alice reads 22 pages every day.
How many pages will Alice read in 9 days?

P10	
a.	
b.	

Part 11

a. 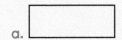 b.  c. d. e. f.

1. Write the letter of each shape that is a triangle.

2. Write the letter of each shape that is a rectangle.

3. Write the letter of each shape that is a hexagon.

4. Write the letter of each shape that is a quadrilateral.

5. Write the letter of each shape that is a square.

P11		
1.	2.	3.
4.	5.	

Lesson 102

Part 1

a. On Monday, the girls played for 1 hour and 5 minutes.
On Tuesday, the girls played for 2 hours and 15 minutes.
How much longer did the girls play on Tuesday
than on Monday?

b. Last night, Alex slept for 8 hours.
Kevin slept for 9 hours and 20 minutes.
How much longer did Kevin sleep than Alex slept last night?

P1	
a.	⟶
b.	⟶

Part 2

P2						
a.	4 8	▆	b. 6 1	▆	c. 3 9	▆
	× 2	× ▆	× 5	× ▆	× 4	× ▆

Independent Work

Part 3 Find the area of each rectangle. Be sure to write the whole unit name.

a.

4 in.

6 in.

b.

5 ft

3 ft

c.

3 cm

3 cm

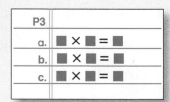

P3	
a.	■ × ■ = ■
b.	■ × ■ = ■
c.	■ × ■ = ■

Part 4 Copy and work each problem.

a. $4 \times 4 \times 2 =$

b. $3 \times 5 \times 4 =$

P4	
a.	■ × ■ × ■ = ■ b. ■ × ■ × ■ = ■

262 Lesson 102

Connecting Math Concepts

Lesson 102

Part 5 Copy and work each problem.

a.   32
    × 9

b.   54
    × 6

c.   24
    × 3

P5			
a. ■	b. ■	c. ■	
■	■	■	
■	■	■	

Part 6 For each item, make the number family and work the problem.

a. The cherry tree was 9 feet tall.
   The fir tree was 4 times as tall as the cherry tree.
   How tall was the fir tree?

b. The green snake was 120 centimeters long.
   The brown snake was 3 times as long as the green snake.
   How many centimeters long was the brown snake?

c. There are 9 dogs and 45 cats in the pet store.
   The number of cats is how many times the number of dogs
   in the pet store?

P6	
a.	
b.	
c.	

Part 7 Copy each item and write the sign >, <, or =.

P7			
a. $\dfrac{4}{4}$ ■ $\dfrac{3}{2}$	b. $\dfrac{8}{10}$ ■ $\dfrac{5}{5}$	c. $\dfrac{6}{6}$ ■ $\dfrac{4}{4}$	

Part 8 Write each number.

a. three thousand thirty

b. three thousand three hundred

c. three thousand three

d. three thousand

P8			
a.	b.	c.	d.

Lesson 102

Part 9 Copy and work each problem.

a. $9\overline{)3\ 6\ 9}$ b. $5\overline{)3\ 5\ 0}$ c. $2\overline{)6\ 2}$

P9			
a.		b.	c.

Part 10 For each equation, write the missing number.

a. $2 \times \blacksquare = 18$ b. $4 \times \blacksquare = 24$ c. $3 \times \blacksquare = 15$

P10			
a.		b.	c.

Part 11 Copy each fraction and write the whole number it equals.

P11			
a. $\dfrac{6}{2}=$		b. $\dfrac{10}{5}=$	c. $\dfrac{9}{3}=$

Lesson 103

Part 1

a. How many pounds did each truck weigh?

b. How many miles did they run each day?

c. How many pages did each book have?

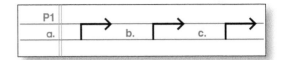

Independent Work

Part 2 Copy and work each problem. Next to the problem, work an estimation problem with the top number rounded to the nearest ten.

P2							
a.	3 8		▆	b.	6 1		▆
	× 5	×	▆		× 4	×	▆

Part 3 For each problem, make a number family and work the problem. Remember to write the dots.

a. Peter reads a book for 1 hour and 5 minutes.
Alex reads a book for 2 hours and 30 minutes.
How much longer does Alex read than Peter reads?

b. Maria worked for 6 hours.
Janice worked for 20 minutes more than Maria worked.
How long did Janice work?

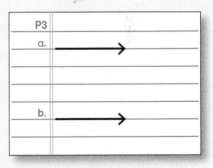

Part 4 Find the area of each rectangle.

a.

3 in.

9 in.

b.

4 ft

5 ft

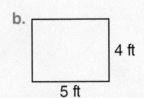

Lesson 103

Part 5 Work the column problem for each item.

a. What is 4 dollars minus $1.70?

b. What is 25 dollars minus $12.30?

P5		
a. ■		b. ■
■		■
■		■

Part 6 Copy and work each problem.

a. $5 \times 3 \times 2 =$ b. $5 \times 9 \times 3 =$

P6	
a. ■ × ■ × ■ = ■	b. ■ × ■ × ■ = ■

Part 7 For each item, make the number family and work the problem.

a. Antonio weighs 48 pounds.
   His father weighs 5 times as much as Antonio.
   How many pounds does his father weigh?

b. Gary read 3 times as many pages as Alex read.
   Gary read 129 pages.
   How many pages did Alex read?

c. Kevin has 9 blue marbles and 27 red marbles.
   Kevin has how many times as many red marbles as
   blue marbles?

P7	
a.	
b.	
c.	

Lesson 103

Part 8 Write the fraction for each picture.

a.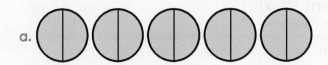

b.

c.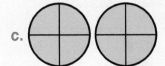

d.

P8				
a. ▪ ― ▪	b. ▪ ― ▪	c. ▪ ― ▪	d. ▪ ― ▪

Part 9 For each item, work the column problem to find the missing number.

a. 80 − ▪ = 20   b. ▪ − 70 = 10   c. 20 + ▪ = 30

P9			
a. ▪ ― ▪	b. ▪ ― ▪	c. ▪ ― ▪

Lesson

Independent Work

Part 1 Copy and work each problem. Next to the problem, work an estimation problem with the top number rounded to the nearest ten.

P1					
a.	4 9	▨	b.	6 2	▨
	× 2	× ▨		× 3	× ▨

Part 2 Work a column problem for each item. Remember to write the dots.

a. The movie started at 5:15.
The movie was 2 hours long.
What time did the movie end?

b. Tina started walking at 8:15.
She walked for 1 hour and 5 minutes.
What time did Tina stop walking?

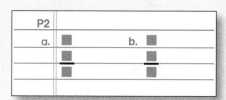

Part 3 Write the fraction for each picture.

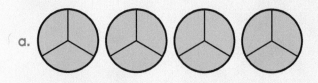

a.

b.

c.

d.

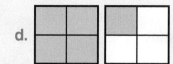

P3							
a.	▪	b. ▪	c. ▪	d. ▪
	▪	▪	▪	▪

Lesson 104

Part 4 Copy and work each problem.

a. 4)3 6 8   b. 5)3 5 0   c. 3)1 2 6

P4			
a.	⌐	b. ⌐	c. ⌐

Part 5 Copy each item and write the sign >, <, or =.

P5			
a. $\dfrac{4}{4}$ ■ $\dfrac{6}{8}$	b. $\dfrac{5}{5}$ ■ $\dfrac{9}{9}$	c. $\dfrac{3}{2}$ ■ $\dfrac{6}{6}$	

Part 6 For each item, make the number family and work the problem. Write the unit name in the answer.

a. We bought 4 pizzas.
We cut each pizza into 6 slices.
How many slices did we have?

b. We put 4 new tires on each car.
We put 48 tires on the cars.
How many cars were there?

c. The red truck weighs 5240 pounds.
The blue truck weighs 5170 pounds.
How much less does the blue truck weigh than the red truck?

P6	
a.	
b.	
c.	

Lesson 104

Part 7 Find the area of each rectangle.

a.

3 ft

7 ft

b.

4 in.

9 in.

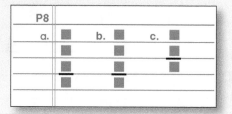

P7	
a.	■ × ■ = ■
b.	■ × ■ = ■

Part 8 Work column problems to answer questions about the table.

This table shows how many hours each person worked during three months.

	October	November	December
Marcos	50	100	150
Lee	75	150	150
Alicia	100	175	150

P8			
a. ■	b. ■	c. ■	
■	■	■	
■	■	■	
■	■		

a. How many hours did the three people work in November?

b. How many hours did Lee work in the three months?

c. In October, how many more hours did Lee work than Marcos worked?

Lesson 105

a.

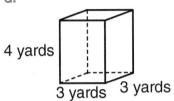

4 yards

3 yards   3 yards

b.

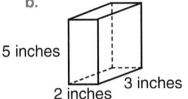

5 inches

2 inches   3 inches

P1	
a.	
b.	

Part 2

a. There were 4 children.
All the children had the same number of stickers.
There were 88 stickers.
How many stickers did each child have?

b. For 9 days, Jan drove the same distance.
She drove a total of 360 miles.
How many miles did she drive each day?

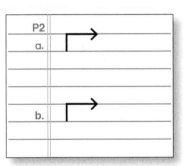

Independent Work

Part 3 | Work a column problem for each item. Remember to write the dollar sign and dot.

a. Jason had 2 dollars.
Thomas had 45 cents.
Robert had 9 cents.
How much money did the boys have in all?

b. Ellen had 7 dollars.
Rhonda had 95 cents.
Sonya had 30 cents.
How much money did the girls have in all?

Part 4 | Copy and work each problem.

a.
```
  342
+  68
```

b.
```
  540
-  94
```

c.
```
  266
+  40
```

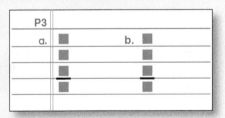

Lesson 105

Part 5 For each equation, write the missing number.

a. $5 \times \blacksquare = 35$ b. $9 \times \blacksquare = 36$ c. $4 \times \blacksquare = 28$

P5		
a.	b.	c.

Part 6 For each item, make the number family and work the problem. Write the unit name in the answer.

a. We watch 4 movies every week.
   How many movies will we watch in 12 weeks?

b. There were lots of people on the train.
   24 more people got on the train.
   Now there are 196 people on the train.
   How many people did the train start with?

c. There are 4 rooms on each floor.
   There are 11 floors.
   How many rooms are there?

P6	
a.	
b.	
c.	

Part 7 Copy each item and write the sign >, <, or =.

P7		
a.	$9 \times 4 \ \blacksquare \ 30 + 5$	b. $40 + 10 \ \blacksquare \ 10 \times 5$

Part 8 Copy and work each problem. Next to the problem, work an estimation problem with the top number rounded to the nearest ten.

P8				
a.	$\begin{array}{r} 4\ 1 \\ \times\ \ 9 \\ \hline \end{array}$	$\blacksquare$ $\times \blacksquare$	b. $\begin{array}{r} 4\ 9 \\ \times\ \ 2 \\ \hline \end{array}$	$\blacksquare$ $\times \blacksquare$

Connecting Math Concepts

Lesson 105

Part 9 Answer the questions about the graph.

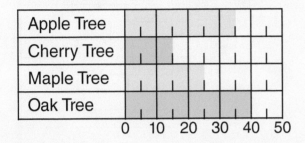

This graph shows how many kinds of trees are in the park.

Apple Tree					
Cherry Tree					
Maple Tree					
Oak Tree					

0 10 20 30 40 50

P9			
a.	b.	c. ▪
		▬
		▪

a. How many cherry trees are in the park?

b. How many oak trees are in the park?

c. How many more apple trees are in the park than maple trees?

Part 10 Write the place-value addition fact for each number.

a. 1254   b. 8114

P10			
a. ▪		b. ▪
▪		▪
▪		▪
▬		▬
▪		▪

Part 11 Work a column problem for each item. Remember to write the dots.

a. The train to New York left at 4 o'clock.
   The trip took 2 hours and 5 minutes.
   When did the train arrive in New York?

b. It started snowing at 3:20.
   It snowed for 30 minutes.
   At what time did it stop snowing?

P11			
a. ▪		b. ▪
▬		▬
▪		▪

Lesson 106

Part 1

a. The garden had 5 bushes.
All the bushes had the same number of flowers.
There were 100 flowers.
How many flowers were on each bush?

b. All the pizzas were on sale for the same price.
There were 9 pizzas.
The total cost for all the pizzas was 99 dollars.
How many dollars did each pizza cost?

c. All the rooms in the motel were the same.
There were 9 rooms and 27 doors.
How many doors were in each room?

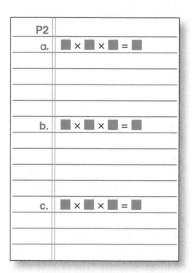

Part 2

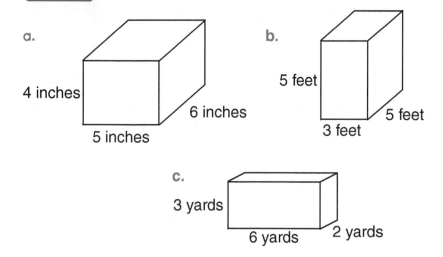

Independent Work

| **Part 3** | Work a column problem for each item. Remember to write the dollar sign and dot. |

a. Ken had 11 dollars.
Mario had 45 cents.
Kevin had 12 dollars.
How much money did the boys have in all?

b. Brenda had 50 cents.
Rhonda had 4 dollars.
Alice had 7 cents.
How much money did the girls have in all?

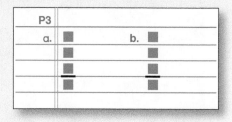

Connecting Math Concepts

Lesson 106

Part 4 Copy and work each problem.

a. $9\overline{)450}$ b. $2\overline{)184}$ c. $4\overline{)208}$

P4					
a.		b.		c.	

Part 5 For each item, make the number family and work the problem. Write the unit name in the answer.

a. The store had lots of apples at the beginning of the day. The store sold 145 apples during the day. The store ended up with 146 apples. How many apples did the store have at the beginning of the day?

b. Every room has 5 windows. There are 30 windows. How many rooms are there?

c. A boy had $4.20 in the bank. He put some more money into the bank. Now he has $7.50 in the bank. How much money did he put into the bank?

d. We bought 5 pizzas. Each pizza had 10 slices. How many slices of pizza did we have?

P5	
a.	
b.	
c.	
d.	

Part 6 Copy each fraction and write the whole number it equals.

P6	
a. $\dfrac{6}{2} =$ b. $\dfrac{8}{4} =$ c. $\dfrac{40}{10} =$	

Lesson 106

Part 7 | Work a column problem for each item. Remember to write the dots.

a. The movie started at 2 o'clock.
   The movie was 1 hour and 15 minutes long.
   What time did the movie end?

b. Mary started walking at 5:30.
   She walked for 1 hour and 5 minutes.
   What time did Mary stop walking?

P7			
a.	■	b.	■
	■		**■**
	■		■

Part 8 | Copy each item and write the sign >, <, or =.

P8		
a.	4 × 4 ■ 30 − 10	b. 5 × 6 ■ 20 + 9

Part 9

a. Find the perimeter of this rectangle.

b. Find the area of this rectangle.

9 ft

3 ft

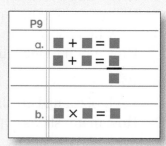

Part 10 | Copy each problem and write the answer.

P10			
a.	$\dfrac{5}{9} + \dfrac{3}{9} =$ ―	b. $\dfrac{4}{7} - \dfrac{1}{7} =$ ―	c. $\dfrac{5}{8} - \dfrac{2}{8} =$ ―

Lesson 106

Part 11 Answer the questions about the graph.

This graph shows how many boys played different sports.

Baseball	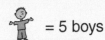
Football	👦 👦 👦 👦
Soccer	👦 👦 👦 👦 👦
Basketball	👦 👦 👦

👦 = 5 boys

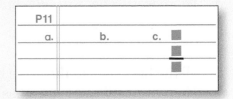

P11			
a.	b.	c.	■
			■ ▬
			■

a. How many boys played football?

b. Which sport did 5 boys play?

c. How many more boys played basketball than baseball?

Lesson 107

Part 1

a. There were 5 pens in each box.
There were 15 boxes.
How many pens were there?

b. All the baskets had the same number of eggs.
There were 40 eggs and 5 baskets.
How many eggs were in each basket?

c. Each truck had 12 wheels.
There were 4 trucks.
How many wheels were there?

d. All the teams had the same number of players.
There were 6 teams and 24 players.
How many players were on each team?

P1	
a.	→
b.	→
c.	→
d.	→

Part 2

a.

b.

c.

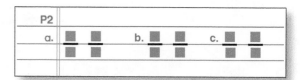

P2				
a. ▪ ▪		b. ▪ ▪	c. ▪ ▪
▪ ▪		▪ ▪	▪ ▪

Part 3

a.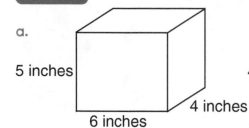

5 inches

6 inches

4 inches

b.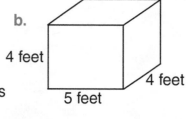

4 feet

5 feet

4 feet

c.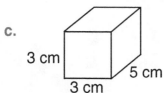

3 cm

3 cm

5 cm

P3	
a.	▪ × ▪ × ▪ = ▪
b.	▪ × ▪ × ▪ = ▪
c.	▪ × ▪ × ▪ = ▪

Lesson

Part 4 For each item, make the number family and work the problem.

a. Nancy read 5 times as many pages as Jennifer read.
   Nancy read 150 pages.
   How many pages did Jennifer read?

b. Elisa walked 24 miles.
   James walked 4 miles.
   Elisa walked how many times the number of miles
   that James walked?

c. The cat weighed 14 pounds.
   The dog weighed 3 times as much as the cat.
   How many pounds did the dog weigh?

P4	
a.	
b.	
c.	

Part 5 Copy and work each problem.

a. 4⟌2 4 8   b. 2⟌1 4 0   c. 3⟌6 9 3

P5			
a.	b.	c.	

Part 6 Write the fraction for each picture.

a.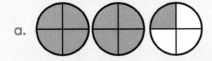

b. (two square figures, one fully shaded, one half shaded)

c.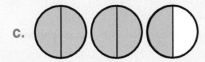

d. (circle divided into four parts)

P6				
a. ▪	b. ▪	c. ▪	d. ▪	
▪	▪	▪	▪	

Part 7 Work a column problem for each item. Remember to write the dots.

a. James left his house at 8:15.
   It took James 35 minutes to walk to school.
   What time did he arrive at school?

b. The train to Washington left at 8 o'clock.
   The trip took 2 hours and 5 minutes.
   What time did the train arrive in New York?

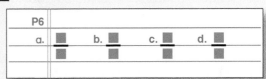

Lesson 107

Part 8

a. Find the perimeter of this rectangle.

b. Find the area of this rectangle.

6 in.

2 in.

P8	
a.	■ + ■ = ■
	■ + ■ = ■
	■
b.	■ × ■ = ■

Part 9 Write the missing number for each fact.

a. 1 pound = ■ ounces

b. 1 minute = ■ seconds

c. 1 dime = ■ cents

d. 1 foot = ■ inches

e. 1 hour = ■ minutes

f. 1 yard = ■ feet

g. 1 week = ■ days

h. 1 gallon = ■ quarts

i. 1 day = ■ hours

j. 1 year = ■ months

P9				
a.	b.	c.	d.	e.
f.	g.	h.	i.	j.

Part 10 Work the column problem for each item.

a. What is 8 dollars minus $3.40?

b. What is 24 dollars minus $11.30?

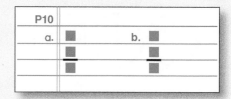

Part 11 Copy and work each problem.

a. 49
× 3

b. 640
× 2

c. 310
× 4

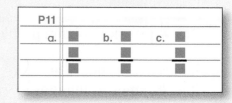

Lesson 108

Part 1

a. There were 8 buildings.
All the buildings had the same number of rooms.
There were 32 rooms.
How many rooms were in each building?

b. There were 8 players on each team.
There were 20 teams.
How many players were there?

c. There were 10 buckets.
Each bucket had 2 gallons of water.
How many gallons of water were there?

d. All the trucks had the same number of wheels.
There were 24 wheels and 3 trucks.
How many wheels were on each truck?

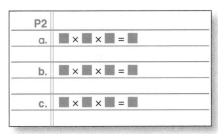

Part 2

a.

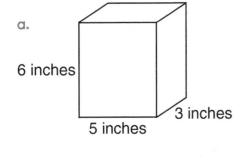

6 inches
5 inches
3 inches

b.

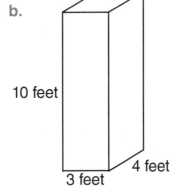

10 feet
3 feet
4 feet

c.

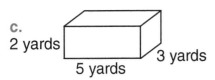

2 yards
5 yards
3 yards

P2	
a.	■ × ■ × ■ = ■
b.	■ × ■ × ■ = ■
c.	■ × ■ × ■ = ■

Independent Work

Part 3 Copy each item and write the sign >, <, or =.

P3			
a. $\dfrac{4}{5}$ ■ $\dfrac{2}{2}$	b. $\dfrac{3}{3}$ ■ $\dfrac{7}{9}$	c. $\dfrac{4}{4}$ ■ $\dfrac{6}{6}$	

Lesson 108

Part 4 | For each item, make the number family and work the problem.
Write the unit name in the answer.

a. The train started with 145 passengers.
Some more passengers got on the train.
The train ended up with 170 passengers.
How many more passengers got on the train?

b. The store started with lots of eggs.
The store sold 140 eggs.
The store ended up with 250 eggs.
How many eggs did the store start with?

c. Sandra did 152 math problems.
Sandra did 28 more math problems than Alice did.
How many math problems did Alice do?

P4	
a.	
b.	
c.	

Part 5 | Answer the questions about the graph.

This graph shows how many days it rained each month.

October	
November	
December	
January	

0 2 4 6 8 10

P5		
a.	b.	c.

a. How many days did it rain in December?

b. In which month did it rain 5 days?

c. How many more days did it rain in January than October?

Part 6 | Work a column problem. Remember to write the dollar sign and dot.

a. Jeremy had 5 dollars.
Elton had 40 cents.
Robert had 5 cents.
How much money did the boys have in all?

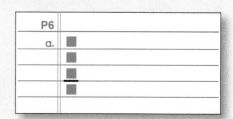

Part 7 | Copy and work each problem.

a. 299
  + 55

b. 320
  −150

c. 1470
  − 85

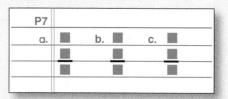

Lesson 108

Part 8 Write the missing number for each fact.

a. 1 year = ■ months   f. 1 hour = ■ minutes

b. 1 day = ■ hours   g. 1 foot = ■ inches

c. 1 gallon = ■ quarts   h. 1 dime = ■ cents

d. 1 week = ■ days   i. 1 minute = ■ seconds

e. 1 yard = ■ feet   j. 1 pound = ■ ounces

P8					
a.		b.	c.	d.	e.
f.		g.	h.	i.	j.

Part 9 Copy and work each problem. Next to the problem, work an estimation problem with the top number rounded to the nearest ten.

P9			
a.	3 2 ■	b.	4 8 ■
	× 7 × ■		× 5 × ■

Part 10 Find the area of each rectangle. Be sure to write the whole unit name.

a.
2 in.
3 in.

b.
4 ft
3 ft

P10	
a.	■ × ■ = ■
b.	■ × ■ = ■

Lesson 109

Part 1

a. 1 yard equals 3 feet.

b. 1 nickel equals 5 cents.

c. 1 gallon equals 4 quarts

d. 1 pound equals 16 ounces.

P1			
a.	⌐→	b.	⌐→
c.	⌐→	d.	⌐→

Independent Work

Part 2 For each item, make the number family and work the problem. Write the unit name in the answer.

a. There were 4 books.
All the books had the same number of pages.
There were 480 pages in all.
How many pages did each book have?

b. Each monkey ate 3 bananas.
The monkeys ate 21 bananas in all.
How many monkeys were there?

c. All the trains had the same number of passengers.
There were 3 trains and 150 passengers.
How many passengers were on each train?

d. There were 15 fish in each lake.
There were 4 lakes.
How many fish were there?

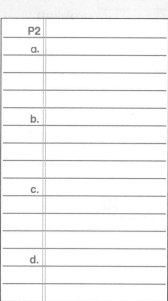

P2	
a.	
b.	
c.	
d.	

Part 3 Copy and work each problem.

a.   3 4 8
    + 6 8

b.   6 2 0 0
    − 4 1 9 0

c.   3 5 2
    − 1 7 7

P3					
a.	■	b.	■	c.	■
	■		■		■
	■		■		■

Connecting Math Concepts

Lesson 109

Part 4 | Write the place-value addition fact for each number.

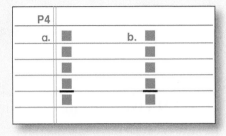

a. 3285   b. 4214

Part 5 | For each item, make the number family and work the problem.

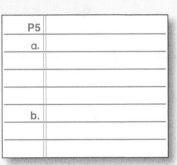

a. The boy had some money.
   He spent $4.50.
   He ended up with $7.00
   How much money did he start with?

b. The shirt cost $9.50.
   The cap cost $1.50 less than the shirt cost.
   How much did the cap cost?

Part 6 | Write the answer to each problem.

a. 30 × 4   b. 9 × 50   c. 2 × 80   d. 40 × 6

Part 7 | Copy each problem and write the answer.

a. $\frac{3}{5} + \frac{1}{5} =$ ___   b. $\frac{4}{9} + \frac{3}{9} =$ ___   c. $\frac{6}{7} - \frac{2}{7} =$ ___

Lesson 109

Part 8 Copy and work each problem.

a. $9 \times 3 \times 5 =$　　　　b. $4 \times 4 \times 2 =$

P8	
a.	■ × ■ × ■ = ■　　b. ■ × ■ × ■ = ■

Part 9 Write the missing number for each number family.

a. $9 \longrightarrow 81$　　f. $4 \longrightarrow 24$　　k. $\_\_\_ \overset{8}{\longrightarrow} 72$

b. $9 \overset{6}{\longrightarrow} \_\_\_$　　g. $9 \overset{8}{\longrightarrow} \_\_\_$　　l. $9 \overset{7}{\longrightarrow} \_\_\_$

c. $\_\_\_ \overset{8}{\longrightarrow} 32$　　h. $\_\_\_ \overset{8}{\longrightarrow} 24$　　m. $9 \longrightarrow 54$

d. $9 \longrightarrow 63$　　i. $9 \longrightarrow 54$　　n. $9 \longrightarrow 72$

e. $3 \overset{8}{\longrightarrow} \_\_\_$　　j. $3 \longrightarrow 21$　　o. $3 \overset{8}{\longrightarrow} \_\_\_$

P9		
a.	f.	k.
b.	g.	l.
c.	h.	m.
d.	i.	n.
e.	j.	o.

Lesson 110

Part 1

a. 1 week equals 7 days.

b. 1 gallon equals 4 quarts.

c. 1 quarter equals 25 cents.

d. 1 pound equals 16 ounces.

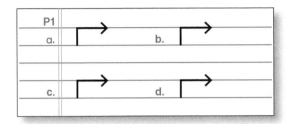

Part 2

a. There were 6 pink blankets and 9 white blankets.
    Each blanket weighed 3 pounds.
    How much did all the blankets weigh?

b. There were 5 vanilla cakes and 6 chocolate cakes.
    Each cake weighed 4 pounds.
    How much did all the cakes weigh?

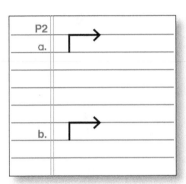

Part 3

a. 5 inches × 2 inches

b. 4 inches, 5 inches, 2 inches

c. 4 feet, 2 feet, 2 feet

d. 2 feet × 2 feet

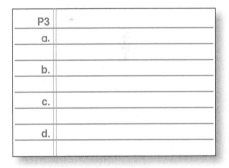

Independent Work

Part 4 Copy and work each problem.

a. 255
   + 95

b. 720
  − 465

c. 3702
  − 902

Lesson 110

Part 5 For each item, make the number family and work the problem. Write the unit name in the answer.

a. There were 6 kittens in each basket.
   There were 11 baskets.
   How many kittens were there?

b. Each cat weighed 12 pounds.
   There were 6 cats.
   How many pounds did the cats weigh altogether?

c. There were 3 mountains.
   All the mountains had the same number of caves.
   There were 150 caves.
   How many caves did each mountain have?

P5	
a.	
b.	
c.	

Part 6 Answer the questions about the graph.

This graph shows how many kinds of animals a pet store sold.

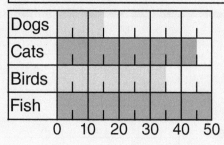

| Dogs | Cats | Birds | Fish |

0 10 20 30 40 50

P6			
a.	b.	c. ■	
		■	
		■	

a. How many birds did the pet store sell?

b. How many fish did the pet store sell?

c. How many more cats than dogs did the pet store sell?

Part 7 Work the column problem for each item.

a. What is 8 dollars minus $1.50?

b. What is 15 dollars minus $3.20?

P7			
a. ■		b. ■	
■		■	
■		■	

Lesson 110

Part 8 Copy and work each problem.

a. $2\overline{)684}$ b. $5\overline{)250}$ c. $3\overline{)156}$

P8			
a.	b.	c.	

Part 9 Copy each item and write the sign >, <, or =.

P9			
a. $\dfrac{7}{7}$ ▪ $\dfrac{3}{3}$	b. $\dfrac{4}{5}$ ▪ $\dfrac{3}{3}$	c. $\dfrac{7}{5}$ ▪ $\dfrac{4}{4}$	

Part 10 Write the missing number for each number family.

a. $9 \xrightarrow{7} \underline{\quad}$

b. $9 \xrightarrow{8} \underline{\quad}$

c. $9 \longrightarrow 54$

d. $9 \longrightarrow 72$

e. $\underline{\quad} \xrightarrow{6} 24$

f. $\underline{\quad} \xrightarrow{9} 27$

g. $3 \longrightarrow 24$

h. $9 \longrightarrow 63$

i. $\underline{\quad} \xrightarrow{8} 32$

j. $4 \longrightarrow 36$

k. $\underline{\quad} \xrightarrow{7} 21$

l. $3 \longrightarrow 18$

m. $9 \longrightarrow 63$

n. $\underline{\quad} \xrightarrow{8} 40$

o. $4 \longrightarrow 32$

P10			
a.	f.	k.	
b.	g.	l.	
c.	h.	m.	
d.	i.	n.	
e.	j.	o.	

Lesson 111

Part 1

a.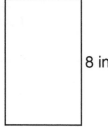
8 in.

5 in.

b.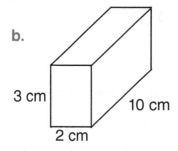
3 cm
10 cm
2 cm

c.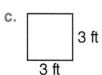
3 ft

3 ft

d.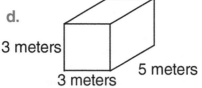
3 meters
5 meters
3 meters

P1	
a.	
b.	
c.	
d.	

Part 2

a. John bought 1 window on Monday and 2 windows on Tuesday.
   Each window cost 40 dollars.
   How much did he pay for the windows?

b. Tom worked 20 hours last week and 10 hours this week.
   He earned 9 dollars for each hour he worked.
   How much did he earn altogether?

P2	⌐→
a.	
b.	⌐→

Independent Work

Part 3 For each item, make a number family and work the problem.

a. James runs 9 kilometers every day.
   How many days will it take James to run 36 kilometers?

b. There are 25 apples in each bag.
   We bought 6 bags of apples.
   How many apples did we buy?

c. All the rooms have the same number of chairs.
   There are 150 chairs and 5 rooms.
   How many chairs are in each room?

P3	
a.	
b.	
c.	

Connecting Math Concepts

Lesson 111

Part 4 Answer the questions about the map.

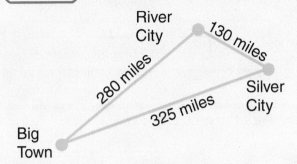

P4			
a.	b.	c. ■	
		■	
		■	

a. Kevin drove from River City to Silver City. How many miles did Kevin drive?

b. Tom drove from Big Town to Silver City. How many miles did Tom drive?

c. How many more miles did Tom drive than Kevin drove?

Part 5 Answer the questions about the graph.

P5			
a.	b.	c. ■	
		■	
		■	

This graph shows how many houses are on each street.

James Street	🏠
Adams Street	🏠 🏠 🏠 🏠 🏠
High Street	🏠 🏠
River Street	🏠 🏠 🏠

🏠 = 5 houses

a. How many houses are on Adams Street?

b. Which street has 5 houses?

c. How many more houses are on River Street than on High Street?

Part 6 Work a column problem. Remember to write the dots.

P6	
a.	■
	■
	■

a. Kelly started reading at 4:15.
She read for 2 hours and 30 minutes.
What time did she stop reading?

Lesson 111

Part 7 Copy and work each problem.

a. $3\overline{)96}$ b. $2\overline{)140}$ c. $4\overline{)128}$

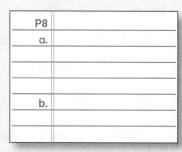

Part 8 For each item, make a number family and work the problem.

a. The truck is 231 kilograms heavier than the car.
   The truck weighs 1495 kilograms.
   How many kilograms does the car weigh?

b. The train started with 190 passengers.
   Some passengers got on the train.
   The train ended up with 245 passengers.
   How many passengers got on the train?

Part 9 Copy each problem and write the answer.

a. $\dfrac{4}{7} + \dfrac{2}{7} =$ ___   b. $\dfrac{3}{5} - \dfrac{1}{5} =$ ___   c. $\dfrac{6}{9} + \dfrac{1}{9} =$ ___

Part 10 For each box, write a fraction to tell how many triangles are big.

a.
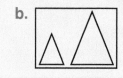

b.

Lesson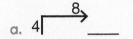

Part 11 Write the missing number for each number family.

a. 4 ⌐8→ ___

b. 4 ⌐→ 28

c. 9 ⌐7→ ___

d. 3 ⌐→ 18

e. 9 ⌐6→ ___

f. 9 ⌐→ 63

g. ___ ⌐8→ 32

h. 9 ⌐8→ ___

i. 3 ⌐8→ ___

j. ___ ⌐6→ 24

k. ___ ⌐6→ 54

l. ___ ⌐4→ 36

m. ___ ⌐8→ 72

n. ___ ⌐8→ 24

o. ___ ⌐7→ 63

P11		
a.	f.	k.
b.	g.	l.
c.	h.	m.
d.	i.	n.
e.	j.	o.

Lesson 112

Part 1

a. John bought 4 short-sleeved shirts and 3 long-sleeved shirts.
Each shirt cost 15 dollars.
How much did he pay for all the shirts?

b. Tom worked 6 hours on Monday and 5 hours on Tuesday.
He earned 8 dollars each hour.
How much did he earn altogether?

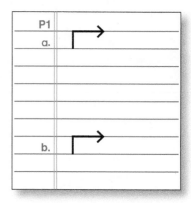

Part 2

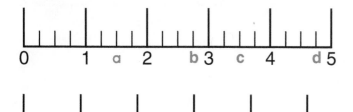

P2			
a.	b.	c.	d.
e.	f.	g.	

Part 3

a.
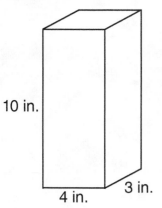

10 in.

4 in.

3 in.

b.
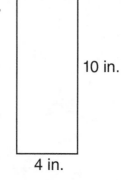

10 in.

4 in.

P3	
a.	
b.	
c.	
d.	

c.

5 cm

6 cm

4 cm

d.

5 cm

6 cm

Lesson 112

> ### Independent Work

Part 4 For each item, make a number family and work the problem.

a. All the trucks have the same number of wheels.
There are 40 wheels on 5 trucks.
How many wheels are on each truck?

b. All the classrooms have the same number of windows.
There are 24 windows and 4 classrooms.
How many windows are in each classroom?

c. Jason does 35 math problems every day.
How many math problems will he do in 7 days?

P4	
a.	
b.	
c.	

Part 5 Copy each problem and write the answer.

P5	
a. $\frac{3}{4} + \frac{2}{4} =$ ___ b. $\frac{4}{7} + \frac{3}{7} =$ ___ c. $\frac{5}{9} - \frac{2}{9} =$ ___	

Part 6 Copy each fraction and write the whole number it equals.

P6	
a. $\frac{40}{10} =$ b. $\frac{10}{2} =$ c. $\frac{6}{3} =$	

Part 7 Write the fraction for each picture.

a.

b.

c.

d.

P7				
a. ▪▬	b. ▪▪	c. ▪▪	d. ▪▪

Lesson 112

Part 8 For each item, make a number family and work the problem.

a. The boy had some money.
He spent $4.50.
He ended up with $7.00.
How much money did he start with?

b. There are 50 fewer boys than girls in the park.
There are 250 boys in the park.
How many girls are in the park?

P8	
a.	
b.	

Part 9 Work column problems to answer questions about the table.

This table shows how many miles each bird flew during 3 months.

	August	September	October
Blue jay	20	30	25
Cardinal	15	35	20
Seagull	50	45	30

P9					
a.	▪	b.	▪	c.	▪
	▪		▪		▪
	▪̲		▪		▪̲
	▪		▪̲		▪
			▪		

a. In October, how many miles did the 3 birds fly altogether?

b. How many miles did the blue jay fly during the 3 months?

c. In August, how many more miles did the seagull fly than the cardinal flew?

Part 10 Write the missing number for each number family.

a. 4 ⟶ 24

f. 4 ⟶ 36

k. __ ⟶ 24 (6)

b. 4 ⟶ __ (8)

g. 9 ⟶ 54

l. __ ⟶ 54 (6)

c. 9 ⟶ 72

h. 4 ⟶ 28

m. __ ⟶ 63 (7)

d. __ ⟶ 35 (7)

i. __ ⟶ 40 (8)

n. __ ⟶ 32 (8)

e. 4 ⟶ 32

j. 9 ⟶ __ (7)

o. 3 ⟶ 18

P10		
a.	f.	k.
b.	g.	l.
c.	h.	m.
d.	i.	n.
e.	j.	o.

Lesson 113

Part 1

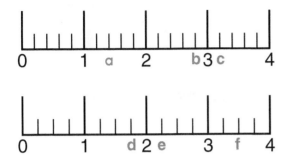

P1			
a.		b.	c.
d.		e.	f.

Part 2

P2							
a.	7 7 7 + 3 3 3	▨ + ▨	▨ + ▨	b.	1 9 4 + 4 1 7	▨ + ▨	▨ + ▨

Independent Work

Part 3 For each problem, first make a number family for the sentence with each. Then work the problem.

a. Each truck had 8 wheels.
   There were 2 red trucks and 2 blue trucks.
   How many wheels were there on all the trucks?

b. Jill bought 2 green shirts and 3 yellow shirts.
   Each shirt cost 9 dollars.
   How many dollars in all did the shirts cost?

P3	
a.	→
b.	→

Lesson 113

Part 4 | Answer the questions about the graph.

This graph shows how many kinds of animals are in the zoo.

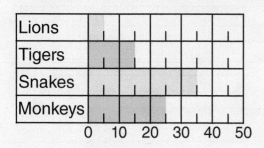

P4		
a. ■		b. ■
■		■
■		■

a. How many more tigers than lions are in the zoo?

b. How many more snakes than monkeys are in the zoo?

Part 5 | Find the volume of each rectangular prism. Write the unit name.

a.
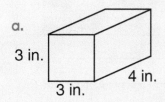

3 in.
3 in.
4 in.

b.
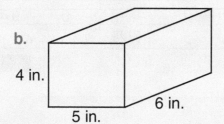

4 in.
5 in.
6 in.

P5	
a.	■ × ■ × ■ = ■
b.	■ × ■ × ■ = ■

Part 6 | Copy and work each problem.

a. 25
  × 6

b. 701
  × 2

c. 32
  × 9

P6				
a. ■		b. ■		c. ■
■		■		■
■		■		■

Part 7 | For each item, make a number family and work the problem.

a. The teacher bought 5 pens.
   All the pens cost the same amount.
   The teacher spent 15 dollars for all the pens.
   How many dollars did each pen cost?

b. We cut each pizza into 9 slices.
   We had 4 pizzas.
   How many slices did we have?

P7	
a.	
b.	

Lesson 113

Part 8 Work the column problem for each item.

a. What is 7 dollars minus $2.60?

b. What is 25 dollars minus $12.90?

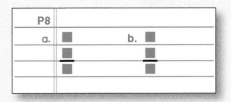

P8		
a. ■		b. ■
■		■
■		■

Part 9 Copy each item and write the sign >, <, or =.

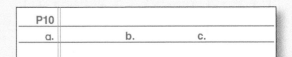

P9	
a. 5×5 ■ $15 + 10$	b. $30 + 6$ ■ 7×5

Part 10 Write the missing number for each row.

a. 40 60 ■ 100

b. 40 50 ■ 70

c. 40 80 ■ 160

P10		
a.	b.	c.

Part 11 Write the missing number for each number family.

a. $9 \xrightarrow{6}$ ___

b. ___ $\xrightarrow{4}$ 28

c. $9 \xrightarrow{\quad}$ 72

d. $3 \xrightarrow{\quad}$ 18

e. $9 \xrightarrow{7}$ ___

f. $3 \xrightarrow{\quad}$ 24

g. $4 \xrightarrow{\quad}$ 24

h. $9 \xrightarrow{8}$ ___

i. $9 \xrightarrow{\quad}$ 63

j. ___ $\xrightarrow{7}$ 28

k. ___ $\xrightarrow{8}$ 72

l. $4 \xrightarrow{\quad}$ 32

m. ___ $\xrightarrow{7}$ 63

n. $4 \xrightarrow{8}$ ___

o. ___ $\xrightarrow{6}$ 54

P11		
a.	f.	k.
b.	g.	l.
c.	h.	m.
d.	i.	n.
e.	j.	o.

Lesson 114

Part 1

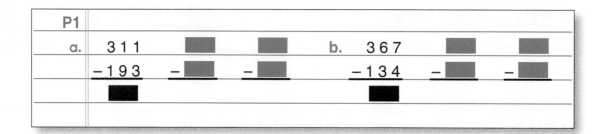

Part 2

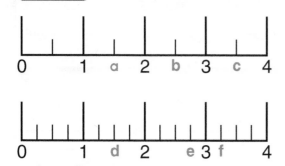

P2			
a.	b.	c.	
d.	e.	f.	

Part 3

a. 1 nickel equals 5 cents.
   How many nickels is 150 cents?

b. 1 day equals 24 hours.
   How many hours is 9 days?

c. 1 quarter equals 25 cents.
   How many cents is 9 quarters?

P3	
a.	→
b.	→
c.	→

Lesson 114

Independent Work

Part 4 For each problem, first make a number family for the sentence with each. Then work the problem.

a. Mary read 3 stories on Monday and 4 stories on Tuesday.
Each story had 12 pages.
How many pages did she read in all?

b. There were 9 birds in each tree.
There were 3 apple trees and 2 cherry trees.
How many birds were there altogether in the trees?

P4	
a.	↱→
b.	↱→

Part 5 Copy each item and write the sign >, <, or =.

P5			
a. $\dfrac{4}{4}$ ▪ $\dfrac{2}{2}$		b. $\dfrac{3}{3}$ ▪ $\dfrac{5}{6}$	c. $\dfrac{4}{4}$ ▪ $\dfrac{3}{2}$

Part 6 Find the volume of each rectangular prism. Write the unit name.

a.

5 feet

4 feet

6 feet

b.

4 feet

5 feet

3 feet

P6	
a.	▪ × ▪ × ▪ = ▪
b.	▪ × ▪ × ▪ = ▪

Part 7 For each item, make a number family and work the problem.

a. The redwood tree is 4 times as tall as the cherry tree.
The redwood tree is 124 feet tall.
How tall is the cherry tree?

b. The red book has 5 times as many pages as the blue book.
The blue book has 20 pages.
How many pages does the red book have?

P7	
a.	
b.	

Lesson 114

Part 8 Work column problems to answer questions about the table.

This table shows how many hours each child played during 3 months.

	January	February	March
Kevin	70	60	50
Marvin	50	45	55
Alex	65	55	55

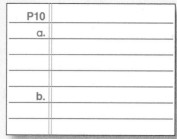

a. In March, how many hours did the 3 children play altogether?

b. How many hours did Marvin play during the 3 months?

c. In February, how many more hours did Alex play than Marvin played?

Part 9 Work the column problem for each item.

a. What is 7 dollars minus $2.60?

b. What is 8 dollars minus $2.50?

Part 10 Work each problem.

a. There are 125 boys and 126 girls in the school. How many children are in the school?

b. There are 60 plates in the kitchen. 32 of the plates are clean. How many of the plates are dirty?

Lesson 114

Part 11 Write the missing number for each number family.

a. $9 \xrightarrow{7}$ ___

b. ___ $\xrightarrow{8} 32$

c. $3 \longrightarrow 18$

d. ___ $\xrightarrow{8} 24$

e. ___ $\xrightarrow{9} 54$

f. $4 \longrightarrow 24$

g. $3 \longrightarrow 24$

h. $4 \xrightarrow{7}$ ___

i. $9 \xrightarrow{8}$ ___

j. $6 \longrightarrow 36$

k. $4 \longrightarrow 28$

l. ___ $\xrightarrow{8} 72$

m. $8 \longrightarrow 32$

n. ___ $\xrightarrow{7} 63$

o. ___ $\xrightarrow{7} 21$

P11		
a.	f.	k.
b.	g.	l.
c.	h.	m.
d.	i.	n.
e.	j.	o.

Lesson 115

Part 1

a. 1 week equals 7 days.
   How many weeks is 28 days?

b. How many days is 5 weeks?

c. 1 foot equals 12 inches.
   How many inches is 4 feet?

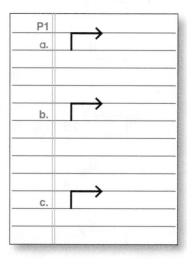

Part 2

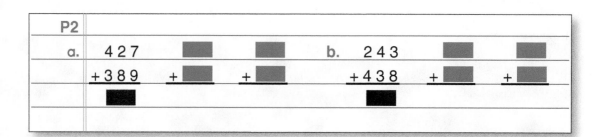

Part 3

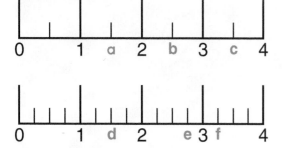

Lesson 115

Part 4 Copy and work each problem.

a. 2⟌864   b. 2⟌126   c. 3⟌90

P4			
a.	⌐	b. ⌐	c. ⌐

Part 5 For each problem, first make a number family for the sentence with each. Then work the problem.

a. Jason bought 3 red shirts and 2 green shirts.
Each shirt cost 9 dollars.
How many dollars did the shirts cost?

b. Chuck earned 3 dollars for every fish he caught.
He caught 4 fish in the morning and 5 fish in the afternoon.
How many dollars did he earn?

P5	
a.	→⌐
b.	→⌐

Part 6 Answer the questions about the graph.

P6			
a.	b.	c. ■
		—
		■

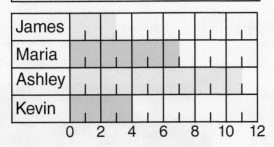

This graph shows how many miles each person ran last week.

a. Who ran more than 8 miles?

b. Who ran less than 4 miles?

c. How many more miles did Maria run than James ran?

Part 7

a. Find the perimeter of this rectangle.

b. Find the area of this rectangle.

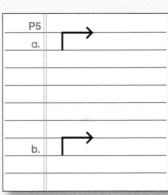

10 in.

4 in.

P7	
a.	■ + ■ = ■
	■ + ■ = ■
	■
b.	■ × ■ = ■

Lesson 115

Part 8 | Work a column problem. Remember to write the dollar sign and dot.

a. Mike had 4 dollars.
   Tom had 83 cents.
   Pedro had 15 dollars.
   How much money did the boys have in all?

P8	
a.	■
	■
	■
	■

Part 9 | Copy each item and write the sign >, <, or =.

P9	
a. $\dfrac{3}{3}$ ■ $\dfrac{6}{8}$	b. $\dfrac{3}{3}$ ■ $\dfrac{5}{5}$	c. $\dfrac{5}{5}$ ■ $\dfrac{4}{3}$

Part 10 | For each item, make a number family and work the problem.

a. All the bags had the same number of apples.
   There were 36 apples in 4 bags.
   How many apples were in each bag?

b. Each street is 45 meters long.
   How many meters long are 6 streets?

c. The green truck weighs 4250 pounds.
   The green truck weighs 150 pounds less than the red truck.
   How much does the red truck weigh?

P10	
a.	
b.	
c.	

Lesson 116

Part 1

a. 5 boys and 4 girls had the same number of pencils.
There were 36 pencils altogether.
How many pencils did each child have?

b. 3 boys and 2 girls read the same number of books.
They read 45 books altogether.
How many books did each child read?

P1	
a.	
b.	

Part 2

a. 1 nickel equals 5 cents.
How many nickels is 150 cents?

b. How many cents is 12 nickels?

c. 1 minute equals 60 seconds.
How many seconds is 5 minutes?

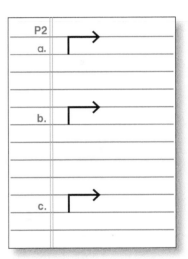

Independent Work

Part 3 For each item, work two estimation problems. For the first problem, round the numbers to the nearest ten. For the second problem, round the numbers to the nearest hundred.

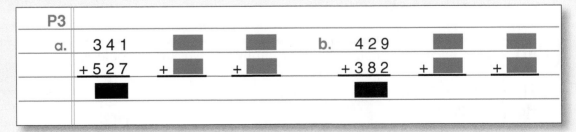

Lesson 116

Part 4 Write the mixed number for each letter.

P4	
a.	b. c.

Part 5 For each item, make the number family and work the problem.

a. Lily ran 400 meters.
   Janice ran 5 times as many meters as Lily ran.
   How many meters did Janice run?

b. Tomás drove 9 kilometers.
   Roberto drove 36 kilometers.
   Roberto drove how many times as far as Tomas drove?

c. Tammy ate 60 grams of peanuts.
   Tammy ate 3 times as many grams of peanuts as Shelly ate.
   How many grams of peanuts did Shelly eat?

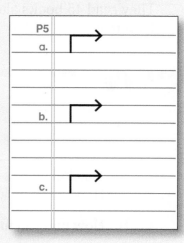

Part 6 Find the volume of each rectangular prism.

a.

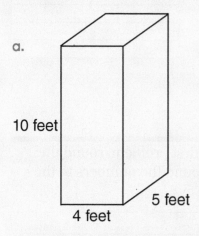

10 feet

4 feet   5 feet

b.

6 in.

5 in.   4 in.

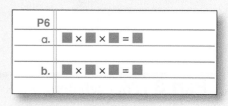

P6	
a.	■ × ■ × ■ = ■
b.	■ × ■ × ■ = ■

Part 7 For each item, make a number family and work the problem.

a. The boy had some money.
   He spent $2.50.
   He ended up with $5.00.
   How much money did he start with?

b. The shirt cost $8.50.
   The shirt cost $1.25 less than the hat cost.
   How much did the hat cost?

P7	
a.	
b.	

Lesson 116

Part 8 For each problem, figure out the hours before 12. Add those hours to the hours after 12.

a. How many hours is it from 10 AM to 5 PM?

b. How many hours is it from 11 AM to 4 PM?

c. How many hours is it from 8 PM to 7 AM?

P8			
a.	■	b. ■	c. ■
	▬	▬	▬
	■	■	■

Part 9 For each equation, write the missing number.

a. $5 \times \blacksquare = 40$ b. $9 \times \blacksquare = 36$ c. $4 \times \blacksquare = 28$

P9		
a.	b.	c.

Part 10 For each box, write a fraction to tell how many faces are smiling.

a.

b.

P10		
a.	▬	b. ▬
	■	■

Lesson 117

Part 1

a. Tom worked 4 hours during the morning and 2 hours during the afternoon.
He earned 24 dollars in all.
He made the same amount all the hours he worked.
How many dollars did he earn each hour?

b. There are 3 bags of red apples and 2 bags of green apples.
There are 30 apples in all.
All the bags have the same number of apples.
How many apples are in each bag?

P1	⟶
a.	
b.	⟶

Part 2

a.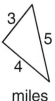
miles

b.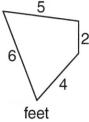
feet

c.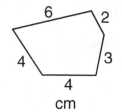
cm

P2	
a.	
b.	
c.	

Part 3

a. 1 nickel equals 5 cents.
How many cents is 7 nickels?

b. How many nickels is 400 cents?

c. 1 quarter equals 25 cents.
How many cents is 7 quarters?

P3	⟶
a.	
b.	⟶
c.	⟶

Part 4

P4			
a. $\dfrac{4}{9} + \dfrac{5}{3} = $ ___	b. $\dfrac{4}{7} + \dfrac{2}{7} = $ ___	c. $\dfrac{5}{8} - \dfrac{1}{8} = $ ___	
d. $\dfrac{3}{5} + \dfrac{1}{5} = $ ___	e. $\dfrac{4}{5} - \dfrac{3}{6} = $ ___	f. $\dfrac{2}{4} - \dfrac{1}{3} = $ ___	

Independent Work

Part 5 For each item, make a number family and work the problem.

a. There are 5 boxes of paper clips.
Each box weighs 125 grams.
How many grams do the 5 boxes of paper clips
weigh altogether?

b. The mountain is 1500 meters high.
The mountain is 700 meters higher than the hill.
How many meters high is the hill?

c. All the coins in a bag are the same weight.
Each coin weighs 5 grams.
The coins weigh 150 grams in all.
How many coins are in the bag?

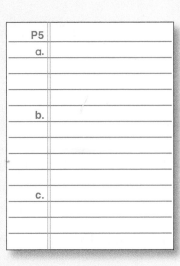

P5	
a.	
b.	
c.	

Part 6 Work two estimation problems. For the first problem, round the numbers to the nearest ten. For the second problem, round the numbers to the nearest hundred.

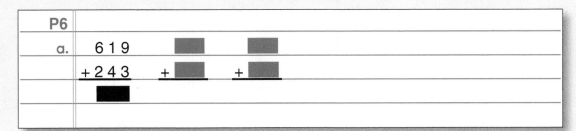

P6			
a.	6 1 9	▬	▬
	+ 2 4 3	+ ▬	+ ▬
	▬		

Lesson 117

Part 7 | Write the mixed number for each letter.

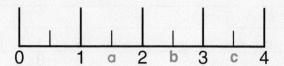

P7			
a.		b.	c.

Part 8

a. Find the perimeter of this rectangle.

b. Find the area of this rectangle.

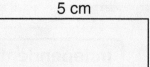

5 cm

2 cm

P8	
a.	■ + ■ = ■
	■ + ■ = ■
	■
b.	■ × ■ = ■

Part 9 | Copy and work each problem.

a. 3)120   b. 9)279   c. 2)86

P9			
a.	⌐	b. ⌐	c. ⌐

Part 10 | Work each problem.

a. What is 9 dollars minus $2.40?

b. What is 15 dollars minus $1.50?

P10		
a.		b.

Part 11 | For each box, write a fraction to tell how many triangles are big.

a.

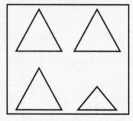

b.

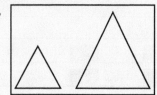

P11		
a.	■ / ■	b. ■ / ■

Lesson 117

Part 12 Write the missing number in each row.

    a. 120   130   ■   150

    b. 120   160   ■   240

    c. 120   150   ■   210

P12		
a.	b.	c.

Part 13 Write the missing number for each number family.

a. 3 →6→ ___

b. ___ →3→ 27

c. ___ →8→ 64

d. 3 →8→ ___

e. 9 →8→ ___

f. 3 →→ 18

g. 9 →→ 63

h. ___ →8→ 24

i. 4 →→ 28

j. 4 →8→ ___

k. 4 →→ 32

l. 7 →→ 63

m. 3 →→ 24

n. ___ →8→ 72

o. ___ →7→ 28

P13		
a.	f.	k.
b.	g.	l.
c.	h.	m.
d.	i.	n.
e.	j.	o.

Lesson 118

Part 1

a. Sam earned the same amount of money for every chair he painted.
He painted 3 new chairs and 2 old chairs.
He earned 15 dollars.
How many dollars did he earn for each chair?

b. Kevin bought 5 red pencils and 3 blue pencils.
All the pencils were the same price.
The pencils cost 80 cents altogether.
How many cents did each pencil cost?

P1	
a.	→
b.	→

Part 2

a.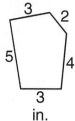
in.

b.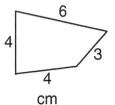
cm

c.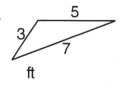
ft

P2	
a.	
b.	
c.	

Part 3

1 yard equals 3 feet
1 gallon equals 4 quarts

a. How many quarts is 10 gallons?

b. How many yards is 12 feet?

c. How many feet is 11 yards?

d. How many gallons is 16 quarts?

P3	
a.	→
b.	→
c.	→
d.	→

Lesson 118

Independent Work

Part 4 Copy and write the answer to the problems you can work.
Don't do anything to the problems you can't work.

P4			
a. $\dfrac{4}{8} + \dfrac{6}{8} =$ ___	b. $\dfrac{4}{12} + \dfrac{9}{5} =$ ___	c. $\dfrac{7}{9} - \dfrac{3}{9} =$ ___
d. $\dfrac{8}{5} + \dfrac{6}{5} =$ ___	e. $\dfrac{5}{7} - \dfrac{4}{7} =$ ___	f. $\dfrac{6}{6} - \dfrac{4}{4} =$ ___

Part 5 Copy and work each problem.

a.
$$\begin{array}{r} 4\,9\,5 \\ \times \quad 3 \\ \hline \end{array}$$

b.
$$\begin{array}{r} 1\,4\,6 \\ \times \quad 5 \\ \hline \end{array}$$

c.
$$\begin{array}{r} 2\,5\,4 \\ \times \quad 6 \\ \hline \end{array}$$

P5			
a. ■	b. ■	c. ■
■	■	■
■	■	■

Part 6 For each item, make a number family and work the problem.

a. A big spider weighed 84 grams.
The big spider weighed 26 grams more than a small spider.
How many grams did the small spider weigh?

b. Jason bought 8 pencils.
Each pencil weighed 25 grams.
How many grams did the pencils weigh in all?

c. There are 4 liters of water in each can.
There are 120 liters of water in all.
How many cans are there?

d. At the beginning of the year, the baby was 54 centimeters tall.
At the end of the year, the baby was 80 centimeters tall.
How many centimeters did the baby grow during the year?

P6	
a.	
b.	
c.	
d.	

Lesson 118

Part 7 | Find the volume of each rectangular prism.

a.

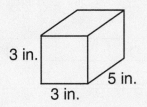

3 in.
3 in.
5 in.

b.

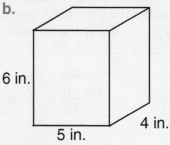

6 in.
5 in.
4 in.

P7	
a.	■ × ■ × ■ = ■
b.	■ × ■ × ■ = ■

Part 8 | Write the mixed number for each letter.

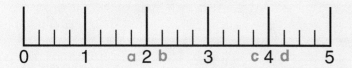

P8				
a.	b.	c.	d.	

Part 9 | Copy and work each problem.

a. 4⟌1 2 8   b. 9⟌3 6 0   c. 3⟌1 5 3

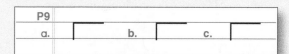

Part 10 | Work two estimation problems. For the first problem, round the numbers to the nearest ten. For the second problem, round the numbers to the nearest hundred.

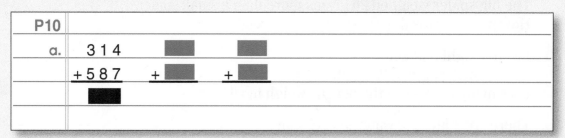

P10			
a.	3 1 4	■	■
	+ 5 8 7	+ ■	+ ■
	■		

Part 11 | Copy each item and write the sign >, <, or =.

P11		
a.	6 × 6 ■ 40 − 10	b. 3 × 3 ■ 4 + 5

Lesson 118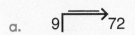

Part 12 Write the missing number for each number family.

a. $9 \xrightarrow{} 72$

b. $9 \xrightarrow{7} \underline{}$

c. $4 \xrightarrow{} 32$

d. $\underline{} \xrightarrow{3} 18$

e. $\underline{} \xrightarrow{9} 27$

f. $6 \xrightarrow{4} \underline{}$

g. $3 \xrightarrow{8} \underline{}$

h. $\underline{} \xrightarrow{3} 18$

i. $\underline{} \xrightarrow{8} 72$

j. $6 \xrightarrow{3} \underline{}$

k. $\underline{} \xrightarrow{8} 24$

l. $9 \xrightarrow{} 63$

m. $\underline{} \xrightarrow{8} 32$

n. $9 \xrightarrow{} 54$

o. $\underline{} \xrightarrow{4} 24$

P12		
a.	f.	k.
b.	g.	l.
c.	h.	m.
d.	i.	n.
e.	j.	o.

Lesson 119

Part 1

a. Dale bought 3 yellow notebooks and 2 white notebooks.
   All the notebooks cost the same amount.
   The notebooks cost 15 dollars in all.
   How many dollars did each notebook cost?

b. Tonya earned 10 dollars each hour she worked.
   She worked 4 hours on Monday and 3 hours on Tuesday.
   How many dollars did she earn?

c. Maria read 3 stories in the morning and 2 stories
   in the afternoon.
   Each story was 15 pages long.
   How many pages in all did Maria read?

d. Jordan bought 2 cotton shirts and 1 wool shirt.
   All the shirts were the same price.
   The shirts cost 27 dollars altogether.
   How many dollars did each shirt cost?

P1	→
a.	→
b.	→
c.	→
d.	→

Part 2

a.

12

18

perimeter = 54 cm

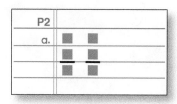

Part 3

a. The store sells 72 books each week.
   About how many books does the store sell
   in 5 weeks?

b. There are 6 windows in each room.
   About how many windows are there in 24 rooms?

c. Jane makes 4 purses every day.
   About how many purses does she make in 48 days?

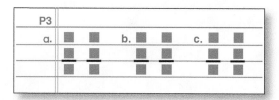

Connecting Math Concepts

Lesson 119

Independent Work

Part 4 | Write the mixed number for each letter.

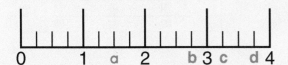

P4				
a.	b.	c.	d.	

Part 5 | Write the numbers for counting by 6s from 36 to 60.

6   12   18   24   30

a.   3__   4__   4__   5__   6__

P5	
a.	▪ ▪ ▪ ▪ ▪

Part 6 | Work a column problem. Remember to write the dollar sign and dot.

a. Jill had 9 cents.
   Ann had 5 dollars.
   Susan had 30 cents.
   How much money did the girls have altogether?

P6	
a.	▪
	▪
	▪
	▬

Part 7 | Copy and write the answer to the problems you can work.
Don't do anything to the problems you can't work.

P7	
a. $\frac{4}{9} + \frac{5}{3} = $ ___ b. $\frac{4}{7} + \frac{7}{8} = $ ___ c. $\frac{5}{6} - \frac{3}{6} = $ ___	
d. $\frac{6}{7} + \frac{4}{7} = $ ___ e. $\frac{7}{10} - \frac{4}{10} = $ ___ f. $\frac{8}{12} - \frac{4}{10} = $ ___	

Part 8 | Work the problem.

a. Ann left her house at 8 o'clock.
   It took her 1 hour and 30 minutes to walk to her office.
   What time did she arrive at her office?

P8	
a.	▪
	▬
	▪

Lesson 119

Part 9 Copy and work each problem.

a. $\begin{array}{r} 1\,3\,4 \\ \times\quad 9 \\ \hline \end{array}$

b. $\begin{array}{r} 8\,6\,5 \\ \times\quad 2 \\ \hline \end{array}$

c. $\begin{array}{r} 2\,5\,6 \\ \times\quad 6 \\ \hline \end{array}$

P9					
a.	■	b.	■	c.	■
	■		■		■
	■		■		■

Part 10 For each item, make a number family and work the problem.

a. There are 16 ounces of juice in each bottle.
   Raymond drank 5 bottles of juice during the week.
   How many ounces of juice did Raymond drink in all?

b. When the day started, the baker had lots of apples.
   The baker used 125 pounds of apples to make pies.
   At the end of the day, the baker had 70 pounds of apples.
   How many pounds of apples did the baker start with?

c. A box of feathers weighed 240 grams.
   Jason put some more feathers in the box.
   Now the box weighs 320 grams.
   How many grams of feathers did Jason put in the box?

d. Jon ran the same number of kilometers each day.
   He went running on 7 days.
   He ran 21 kilometers in all.
   How many kilometers did Jon run each day?

P10	
a.	
b.	
c.	
d.	

Part 11 Write the missing number for each number family.

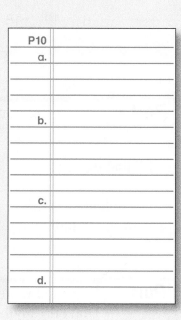

a. $4 \xrightarrow{} 28$

b. $\underline{} \xrightarrow{4} 24$

c. $3 \xrightarrow{} 24$

d. $4 \xrightarrow{8} \underline{}$

e. $\underline{} \xrightarrow{7} 49$

f. $\underline{} \xrightarrow{9} 81$

g. $9 \xrightarrow{} 63$

h. $4 \xrightarrow{7} \underline{}$

i. $9 \xrightarrow{} 72$

j. $\underline{} \xrightarrow{8} 32$

k. $\underline{} \xrightarrow{8} 64$

l. $6 \xrightarrow{4} \underline{}$

m. $\underline{} \xrightarrow{8} 72$

n. $\underline{} \xrightarrow{7} 63$

o. $4 \xrightarrow{} 32$

P11			
a.	f.		k.
b.	g.		l.
c.	h.		m.
d.	i.		n.
e.	j.		o.

Lesson 120

Part 1

> 1 week equals 7 days
> 1 nickel equals 5 cents

a. How many cents is 12 nickels?

b. How many weeks is 21 days?

c. How many nickels is 150 cents?

d. How many days is 4 weeks?

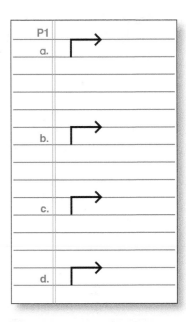

Part 2

a. There are 32 boxes on each shelf.
   There are 8 shelves.
   About how many boxes are there in all?

b. The city had 23 baseball teams.
   There were 9 players on each team.
   About how many players were there altogether?

c. Each classroom had 39 chairs.
   About how many chairs were in 5 classrooms?

Part 3

a.

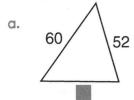

60 52

perimeter = 162 in.

b.

12

9

perimeter = 27 ft

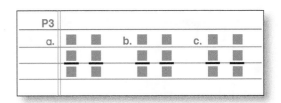

c.

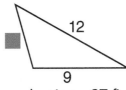

8

10

perimeter = 24 cm

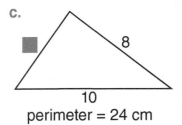

Lesson 120

Part 4

a. Eva earned 4 dollars for each chair she built.
She built 10 chairs on Wednesday and 20 chairs on Thursday.
How many dollars did Eva earn?

b. Jordan read 4 stories in the morning and 5 stories in the afternoon. All the stories had the same number of pages.
He read 90 pages in all.
How many pages did each story have?

c. All the classrooms have the same number of students.
There are 5 classrooms on the first floor and 4 classrooms on the second floor. There are 180 students in all.
How many students are in each classroom?

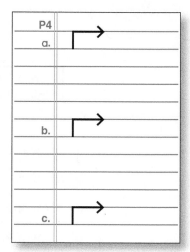

Independent Work

Part 5 | Copy and work each problem.

a. 2 7 5
    ×   3

b. 4 8 6
    ×   2

c. 3 5 2
    ×   4

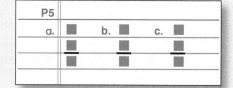

Part 6 | Write the mixed number for each letter.

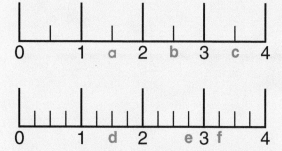

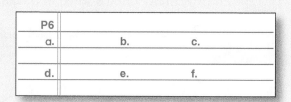

Part 7

a. Find the perimeter of this rectangle.

b. Find the area of this rectangle.

8 cm

3 cm

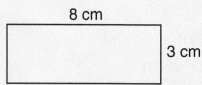

Lesson 120

Part 8 For each item, make a number family and work the problem.

a. A hammer weighed 5 pounds.
A saw weighed 3 times as many pounds as the hammer.
How many pounds did the saw weigh?

b. Rosa ran 75 kilometers.
Tina ran 4 times as many kilometers as Rosa ran.
How many kilometers did Tina run?

P8	
a.	
b.	

Part 9 Copy and write the answer to the problems you can work.
Don't do anything to the problems you can't work.

P9	
a. $\dfrac{7}{3} + \dfrac{2}{3} = $ ___ b. $\dfrac{4}{5} + \dfrac{5}{5} = $ ___ c. $\dfrac{7}{8} - \dfrac{3}{6} = $ ___	

Part 10 Write the numbers for counting by 6s from 36 to 60.

6   12   18   24   30

a. 3__   4__   4__   5__   6__

P10	
a.	■ ■ ■ ■ ■

Part 11 Write the answer to each question.

This graph shows how many days it snowed during four months.

P11		
a.		b.
c.		d.

December	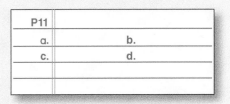
January	
February	
March	

a. How many days did it snow in March?

b. In which month did it snow 2 days?

c. In which month did it snow 4 days?

d. How many more days did it snow in February than in March?

= 2 days

Lesson 120

Part 12 Write the missing number for each number family.

a. $4 \overset{\quad}{\longrightarrow} 32$

b. $9 \overset{7}{\longrightarrow} \underline{\quad}$

c. $9 \overset{\quad}{\longrightarrow} 72$

d. $\underline{\quad} \overset{7}{\longrightarrow} 49$

e. $\underline{\quad} \overset{9}{\longrightarrow} 81$

f. $\underline{\quad} \overset{3}{\longrightarrow} 18$

g. $4 \overset{8}{\longrightarrow} \underline{\quad}$

h. $4 \overset{7}{\longrightarrow} \underline{\quad}$

i. $4 \overset{\quad}{\longrightarrow} 36$

j. $\underline{\quad} \overset{4}{\longrightarrow} 24$

k. $\underline{\quad} \overset{8}{\longrightarrow} 72$

l. $3 \overset{\quad}{\longrightarrow} 24$

m. $\underline{\quad} \overset{7}{\longrightarrow} 63$

n. $\underline{\quad} \overset{8}{\longrightarrow} 64$

o. $4 \overset{\quad}{\longrightarrow} 28$

P12		
a.	f.	k.
b.	g.	l.
c.	h.	m.
d.	i.	n.
e.	j.	o.

Lesson 121

Part 1

a. There are 282 students in each school.
   About how many students are in 5 schools?

b. Jennifer walks for 142 minutes every day.
   About how many minutes will Jennifer walk in 3 days?

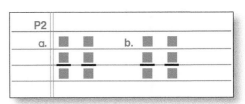

Part 2

a.

perimeter = 50 in.

b. 38
   26

perimeter = 80 in.

Independent Work

Part 3 For each item, make a number family and work the problem.

a. Jason earned 10 dollars each hour he painted the house.
   Jason painted the house for 3 hours during the morning
   and for 5 hours during the afternoon.
   How many dollars did Jason earn?

b. Alejandra bought 5 long-sleeved shirts and 4 short-sleeved shirts.
   The shirts all cost the same number of dollars.
   Alejandra spent 81 dollars altogether for the shirts.
   How many dollars did each shirt cost?

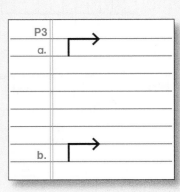

Lesson 121

Part 4 Copy and work each problem.

a.
$$696 \times 2$$

b.
$$70 \times 5$$

c.
$$234 \times 4$$

P4				
a.	■	b. ■	c.	■
	■	■		■
	■	■		■

Part 5 For each item, make a number family and work the problem.

a. 1 pound equals 16 ounces.
How many ounces is 5 pounds?

b. 1 yard equals 3 feet.
How many yards is 150 feet?

c. 1 week equals 7 days.
How many days is 20 weeks?

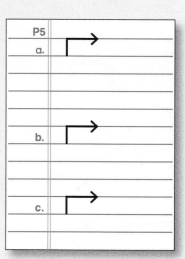

Part 6 Write the mixed number for each letter.

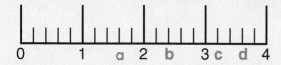

0   1   a 2 b   3 c d 4

P6				
a.		b.	c.	d.

Part 7 Copy and write the answer to the problems you can work.
Don't do anything to the problems you can't work.

P7			
a. $\dfrac{4}{9} + \dfrac{2}{9} =$ ___	b. $\dfrac{5}{7} - \dfrac{3}{4} =$ ___	c. $\dfrac{5}{7} - \dfrac{1}{7} =$ ___	

Connecting Math Concepts

Lesson 121

Part 8 For each item, make the number family and work the problem.

a. All the classes had the same number of students.
   There were 5 classes and 150 students.
   How many students were in each class?

b. Jason does 5 problems every minute.
   How many problems will Jason do in 20 minutes?

P8	
a.	
b.	

Part 9 Copy and work each problem.

a. 4$\overline{)248}$ b. 2$\overline{)180}$ c. 3$\overline{)150}$

P9					
a.		b.		c.	

Part 10 For each item, make a number family and work the problem.

a. The man had some money.
   He spent 35 dollars.
   He ended up with 65 dollars.
   How much money did he start with?

b. The new car cost 8000 dollars.
   The new car cost 4500 dollars more than the old car.
   How much did the old car cost?

P10	
a.	
b.	

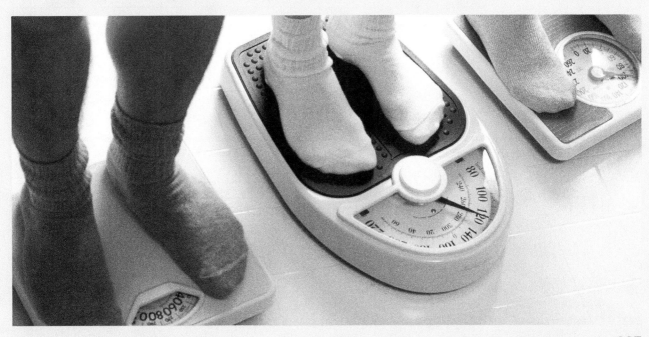

Lesson 122

Part 1

a. A bus started out with 46 passengers.
   7 passengers got off the bus. Then 9 more
   passengers got off the bus.
   How many passengers were still on the bus?

b. The store started out with 45 apples.
   The store sold 8 apples in the morning and
   7 apples in the afternoon.
   How many apples did the store end up with?

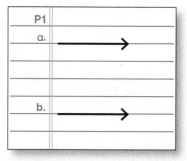

Part 2

a.

perimeter = 15 ft

b.

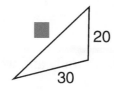

perimeter = 90 cm

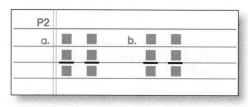

Part 3

a. Janice ran 842 meters on Monday and 538 meters on Tuesday.
   About how many meters did she run altogether?

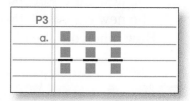

Independent Work

Part 4 Copy and work each problem.

a. 4⟌1 2 0   b. 2⟌8 4   c. 3⟌1 5 6

Lesson 122

Part 5 Work each problem.

a. What is 8 dollars minus $1.60?

b. What is 7 dollars minus $2.30?

P5		
a.		b.

Part 6 Write the mixed number for each letter.

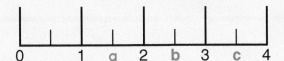

P6			
a.		b.	c.

Part 7 For each item, make a number family and work the problem.

P7	
a.	
b.	
c.	

a. 1 hour equals 60 minutes.
How many minutes is 5 hours?

b. 1 week equals 7 days.
How many weeks is 140 days?

c. 1 nickel equals 5 cents.
How many nickels is 150 cents?

Part 8 For each box, write a fraction to show how many faces are smiling.

a.

b.

P8		
a. ▪		b. ▪
▪		▪

Lesson 122

Part 9 For each item, make a number family and work the problem.

a. The cat weighed 14 pounds.
   The dog weighed 3 times as much as the cat weighed.
   How many pounds did the dog weigh?

b. John has 30 red pencils and 5 blue pencils.
   John has how many times as many red pencils
   as blue pencils?

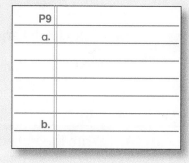

Part 10 Copy and work each problem.

a. $\begin{array}{r} 2\,6\,5 \\ \times\quad 3 \\ \hline \end{array}$
b. $\begin{array}{r} 3\,6 \\ \times\ 2 \\ \hline \end{array}$
c. $\begin{array}{r} 5\,4\,1 \\ \times\quad 9 \\ \hline \end{array}$

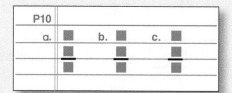

Part 11 Make a number family and work the problem.

a. Lebron wrote 4 stories on Monday and 5 stories on Tuesday.
   Each story was 125 words long.
   How many words did Lebron write?

P11	
a.	

Part 12 Write the missing number for each number family.

a. $9 \xrightarrow{\quad 7\quad} \underline{}$

b. $8 \xrightarrow{\qquad} 64$

c. $4 \xrightarrow{\quad 8\quad} \underline{}$

d. $9 \xrightarrow{\qquad} 81$

e. $4 \xrightarrow{\quad 7\quad} \underline{}$

f. $3 \xrightarrow{\quad 8\quad} \underline{}$

g. $8 \xrightarrow{\qquad} 72$

h. $9 \xrightarrow{\qquad} 63$

i. $\underline{} \xrightarrow{\quad 8\quad} 24$

j. $\underline{} \xrightarrow{\quad 8\quad} 64$

k. $4 \xrightarrow{\qquad} 32$

l. $\underline{} \xrightarrow{\quad 4\quad} 24$

m. $\underline{} \xrightarrow{\quad 8\quad} 72$

n. $\underline{} \xrightarrow{\quad 7\quad} 28$

o. $\underline{} \xrightarrow{\quad 7\quad} 63$

P12		
a.	f.	k.
b.	g.	l.
c.	h.	m.
d.	i.	n.
e.	j.	o.

Lesson 123

Part 1

a. Donna had 144 apples.
She sold 6 apples in the morning and 7 apples in the afternoon.
How many apples did she have left?

b. There were 50 children in the park.
4 boys and 6 girls went home.
How many children were still in the park?

P1	
a.	→
b.	→

Part 2

a. A pizza was divided into 9 equal slices.
The dog ate 2 of those slices.
What fraction of the pizza was left?

b. A field was divided into 10 equal parts.
7 of those parts were mowed.
What fraction of the field was not mowed?

c. A board was 5 feet long.
A worker cut 3 feet from the board.
What fraction of the board was left?

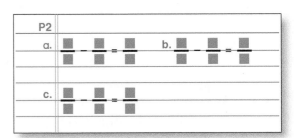

Part 3

a.

5 in.

4 in.

b.

10 in.

2 in.

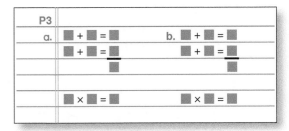

Lesson

Independent Work

Part 4 | Work an estimation problem rounding both numbers to the nearest hundred.

a. Koby ran 870 meters.
Tomás ran 540 meters.
About how many meters did the boys run altogether?

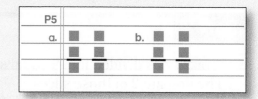

Part 5 | For each triangle, figure out the length of the side that is not shown.

a.

8

7

perimeter = 20 in.

b. 17

10

perimeter = 39 in.

P5		
a. ■ ■		b. ■ ■
■ ■		■ ■
■ ■		■ ■

Part 6 | Copy the problems you can work and write the answers.
Don't do anything to the problems you can't work.

P6			
a. $\frac{5}{7} + \frac{3}{4} = $ ___	b. $\frac{7}{9} - \frac{1}{4} = $ ___	c. $\frac{5}{8} + \frac{2}{8} = $ ___
d. $\frac{9}{10} - \frac{2}{10} = $ ___	e. $\frac{4}{7} + \frac{2}{7} = $ ___	f. $\frac{5}{9} + \frac{3}{4} = $ ___

Part 7 | Write the mixed number for each letter.

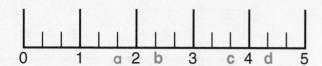

P7			
a.	b.	c.	d.

Lesson 123

Part 8 | Copy and work each problem.

a. $5\overline{)450}$ b. $2\overline{)148}$ c. $4\overline{)208}$

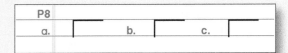

Part 9 | For each item, make a number family and work the problem.

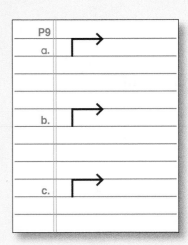

   a. 1 gallon equals 4 quarts.
     How many quarts is 20 gallons?

   b. 1 year equals 12 months.
     How many months is 4 years?

   c. 1 meter equals 100 centimeters.
     How many centimeters is 5 meters?

Part 10 | Copy and work each problem.

a.   243
   ×   6

b.   50
   ×   4

c.   242
   ×   8

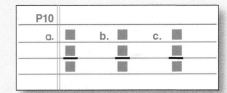

Part 11 | Write an estimation problem rounding both numbers to the nearest hundred.

a. The truck driver drove her truck 571 kilometers
   on Monday and 638 kilometers on Tuesday.
   About how many kilometers did she drive on the two days?

Part 12 | Write the mixed number for each letter.

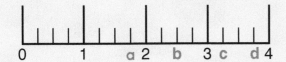

Lesson 123

Part 13 Write the missing number for each number family.

a. 9 ⌐8→ ___

b. 8 ⌐⇒ 64

c. ___ ⌐6→ 18

d. 9 ⌐7→ ___

e. 7 ⌐7→ ___

f. 4 ⌐8→ ___

g. ___ ⌐7→ 49

h. 9 ⌐⇒ 72

i. ___ ⌐8→ 32

j. ___ ⌐9→ 81

k. 3 ⌐⇒ 18

l. ___ ⌐7→ 63

m. ___ ⌐9→ 63

n. 8 ⌐8→ ___

o. 9 ⌐⇒ 72

P13			
a.		f.	k.
b.		g.	l.
c.		h.	m.
d.		i.	n.
e.		j.	o.

Lesson 124

Part 1

a. Pam had 46 marbles.
She gave 9 marbles to her brother and 7 marbles to her sister.
How many marbles did Pam end up with?

b. There were 73 people at the beach.
8 men and 4 women left the beach.
How many people were still at the beach?

P1	
a.	⟶
b.	⟶

Part 2

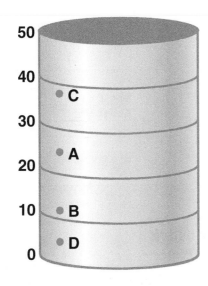

P2				
1.	2.	3.	4.	

1. Which letter shows the tank with about 38 liters of water?

2. Which letter shows the tank with about 5 liters of water?

3. Which letter shows the tank with about 25 liters of water?

4. Which letter shows the tank with about 12 liters of water?

Part 3

a. Mrs. Thomas wants to build a fence around her vegetable garden.
The garden is shaped like a rectangle.
The garden is 22 feet wide and 10 feet long.
How many feet of fence will Mrs. Thomas need?

b. Mr. Jenkins will ride his bike all the way around a path that is shaped like a rectangle.
The path is 11 miles wide and 5 miles long.
If Mr. Jenkins rides his bike all the way around the path, how many miles will he ride?

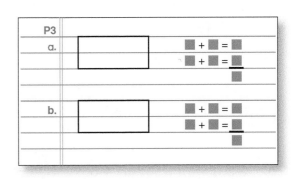

Lesson 124

Part 4

a. A field was divided into 8 equal parts.
   5 of those parts were mowed.
   What's the fraction for the parts that were not mowed?

b. A board was 9 feet long.
   A worker cut 5 feet from the board.
   What fraction of the board was left?

c. A pizza was divided into 4 equal slices.
   The girls ate 3 of those slices.
   What fraction of the pizza was left?

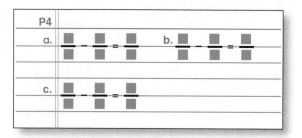

Part 5

a.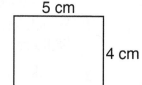
   7 cm
   2 cm

b. 5 cm
   4 cm

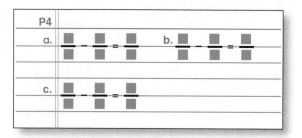

Independent Work

Part 6 For each triangle, figure out the length of the side that is not shown.

a.
14
13
perimeter = 38 in.

b.
24
18
perimeter = 54 in.

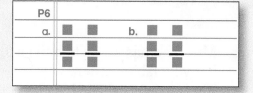

Part 7 Copy and work each problem.

a. 2 3 0
    ×   9

b. 1 4 5
    ×   3

c. 6 0
    × 4

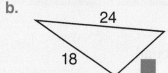

Connecting Math Concepts

Lesson 124

Part 8 Work the problem.

a. Elena had 90 cents.
Rose had 4 dollars.
Jenny had 7 cents.
How much money did the girls have altogether?

P8	
a.	

Part 9 For each box, write a fraction to show how many triangles are big.

a.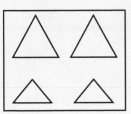

b.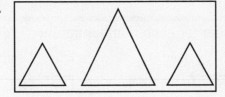

P9			
a.	▪ / ▪	b.	▪ / ▪

Part 10 Copy and work each problem.

a. $2\overline{)6\,8}$ b. $5\overline{)3\,5\,0}$ c. $3\overline{)1\,2\,9}$

P10			
a.	⌐	b. ⌐	c. ⌐

Part 11 Answer the questions about the graph.

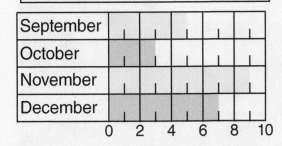

This graph shows how many days it rained each month.

P11			
a.	▪ ▪ ▪	b.	▪ ▪ ▪

a. How many more days did it rain in November than in October?

b. How many more days did it rain in December than in October?

Part 12 For each rectangle, write a times problem and figure out the number of squares.

a.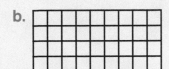

b.

P12		
a.	▪ × ▪ = ▪	b. ▪ × ▪ = ▪

Part 13 Work each problem.

a. Each bag of feathers weighed 50 grams.
How much did 9 bags of feathers weigh?

b. There were 2500 liters of water in the swimming pool.
We put in some more water.
The pool ended up with 2750 liters of water.
How many liters of water did we put into the pool?

P13	
a.	
b.	

Part 14 Write the missing number for each number family.

a. $9 \xrightarrow{\quad} 72$

b. $3 \xrightarrow{8} \underline{\quad}$

c. $\underline{\quad} \xrightarrow{8} 64$

d. $\underline{\quad} \xrightarrow{7} 49$

e. $4 \xrightarrow{8} \underline{\quad}$

f. $9 \xrightarrow{7} \underline{\quad}$

g. $\underline{\quad} \xrightarrow{6} 36$

h. $9 \xrightarrow{8} \underline{\quad}$

i. $4 \xrightarrow{8} \underline{\quad}$

j. $\underline{\quad} \xrightarrow{8} 24$

k. $\underline{\quad} \xrightarrow{6} 24$

l. $\underline{\quad} \xrightarrow{8} 32$

m. $9 \xrightarrow{\quad} 63$

n. $\underline{\quad} \xrightarrow{4} 36$

o. $\underline{\quad} \xrightarrow{8} 72$

P14		
a.	f.	k.
b.	g.	l.
c.	h.	m.
d.	i.	n.
e.	j.	o.

Lesson 125

Part 1

a. Mr. Cruise rode his bike all the way around a park that is shaped like a rectangle.
The park is 10 kilometers wide and 4 kilometers long.
How many kilometers did Mr. Cruise ride?

b. Mrs. Lee walked her dog around a field that is shaped like a rectangle.
The field is 100 meters wide and 40 meters long.
How many meters did Mrs. Lee walk her dog?

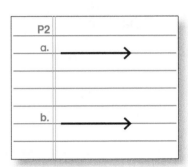

Part 2

a. Linda had 33 dollars.
She spent 4 dollars at the market and 9 dollars at the mall.
How many dollars did she have left?

b. There were 29 birds sitting on a tree branch.
5 robins and 10 bluebirds flew away.
How many birds were left on the tree branch?

Part 3

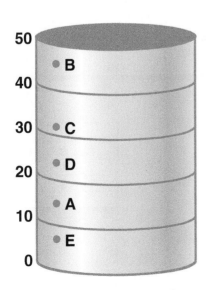

1. Which letter shows the tank with about 7 liters of water?

2. Which letter shows the tank with about 32 liters of water?

3. Which letter shows the tank with about 15 liters of water?

4. Which letter shows the tank with about 24 liters of water?

5. Which letter shows the tank with about 48 liters of water?

Lesson 125

Part 4

a. 4 ft

3 ft

b. 6 ft

2 ft

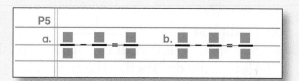

Independent Work

Part 5

For each item, work a fraction problem.

a. A pizza was divided into 7 equal slices.
The boy ate 2 of those slices.
What fraction of the pizza was left?

b. A pie was divided into 8 equal pieces.
5 of the pieces had whipped cream on top.
What fraction of the pie did not have whipped
cream on top?

Part 6

For each triangle, figure out the length of the side that is not shown.

a.

11

8

perimeter = 30 cm

b.

7   9

perimeter = 29 cm

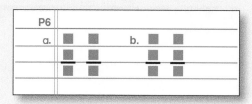

Part 7

Copy and work each problem.

a. 3 2 4
    ×   5

b. 1 8
    × 5

c. 1 3 0
    ×   9

Lesson 125

Part 8 | Copy each item and write the sign >, <, or =.

P8	
a.	6×5 ■ $20 + 10$ b. $25 + 10$ ■ 8×5

Part 9 | Work each problem.

a. The gray horse weighs 54 kilograms more than the brown horse.
The gray horse weighs 456 kilograms.
How many kilograms does the brown horse weigh?

b. The redwood tree is 3 times as tall as the maple tree.
The maple tree is 24 meters tall.
How many meters tall is the redwood tree?

c. Each can has 24 liters of milk.
How many liters of milk are in 5 cans?

P9	
a.	
b.	
c.	

Part 10 | Find the volume of each rectangular prism.

a.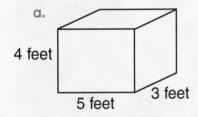
4 feet
5 feet
3 feet

b.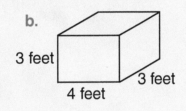
3 feet
4 feet
3 feet

P10	
a.	■ × ■ × ■ = ■
b.	■ × ■ × ■ = ■

Part 11 | Copy and work each problem.

a. 4 3 5
  + 9 0

b. 6 4 0
  − 9 0

c. 5 3 2
  − 9 9

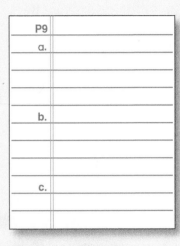

P11			
a. ■	b. ■	c. ■	
■	■	■	
■	■	■	

Lesson 125

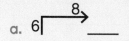

Part 12 Write the missing number for each number family.

a. 6 —8→ ___

b. 9 ═══→ 72

c. 9 —6→ ___

d. 6 ═══→ 48

e. 8 —8→ ___

f. ___ —6→ 54

g. ___ —8→ 24

h. 6 ═══→ 48

i. ___ —8→ 72

j. 6 —6→ ___

k. ___ —8→ 48

l. ___ —9→ 81

m. ___ —8→ 64

n. ___ —6→ 24

o. ___ —6→ 48

P12		
a.	f.	k.
b.	g.	l.
c.	h.	m.
d.	i.	n.
e.	j.	o.

Lesson 126

a. 10 in. | 3 in.

b. 9 in. | 4 in.

P1		
a.	■ + ■ = ■	b. ■ + ■ = ■
	■ + ■ = ■	■ + ■ = ■
	■	■
	■ × ■ = ■	■ × ■ = ■

Independent Work

Part 2 Make a picture of the park. Then work the problem to answer the question.

a. Cindy ran around a park that is shaped like a rectangle.
The park is 100 yards wide and 50 yards long.
How many yards did Cindy run?

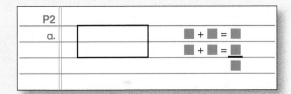

Part 3 Work the problem.

a. The store started the day with 74 bottles of milk.
The store sold 14 bottles of milk in the morning
and 12 bottles of milk in the afternoon.
How many bottles of milk did the store end up with?

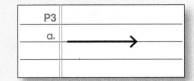

Part 4 Copy and work each problem.

a. 40
  × 9

b. 247
  ×   5

c. 29
  × 2

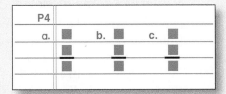

Lesson 126

Part 5 For each item, work a fraction problem.

a. A pie was divided into 7 equal slices.
The dog ate 5 pieces of the pie.
What fraction of the pie was left?

b. A board was 9 feet long.
A worker cut 6 feet from the board.
What fraction of the board was left?

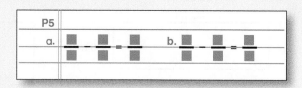

Part 6 Copy and work each problem.

a. 4⟌168   b. 2⟌140   c. 4⟌88

Part 7 Work column problems to answer questions about the table.

This table shows how much money each
person earned during three months.

	May	June	July
James	$150	$125	$150
Alex	$140	$125	$240
Rose	$160	$125	$250

a. How much did the three people earn in June?

b. How much did Alex earn in the three months?

c. In May, how much more money did Rose earn than Alex earned?

Part 8 Work each problem.

a. Kendra had $14.
She earned some money by helping her uncle clean his truck.
Now she has $20.50.
How much money did she earn?

b. 2500 people live in Hill City.
500 more people live in Hill City than in River City.
How many people live in River City?

Lesson 126

Part 9 Write the missing number for each number family.

a. 6 ⌐→ 48

b. ___ ⌐8→ 24

c. 6 ⌐6→ ___

d. ___ ⌐9→ 54

e. ___ ⌐6→ 24

f. ___ ⌐6→ 48

g. 4 ⌐→ 32

h. 6 ⌐8→ ___

i. ___ ⌐8→ 32

j. ___ ⌐3→ 18

k. ___ ⌐6→ 18

l. ___ ⌐6→ 54

m. ___ ⌐4→ 24

n. ___ ⌐3→ 24

o. ___ ⌐8→ 48

P9		
a.	f.	k.
b.	g.	l.
c.	h.	m.
d.	i.	n.
e.	j.	o.

Lesson

Part 1

P1					
1.	2.	3.	4.	5.	6.

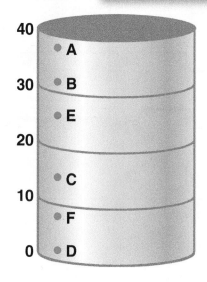

1. Which letter shows the tank with about 8 liters of water?

2. Which letter shows the tank with about 32 liters of water?

3. Which letter shows the tank with about 15 liters of water?

4. Which letter shows the tank with about 38 liters of water?

5. Which letter shows the tank with about 2 liters of water?

6. Which letter shows the tank with about 26 liters of water?

Independent Work

Part 2 Make a picture of the yard. Then work the problem to answer the question.

a. Mr. James wants to build a fence around his garden.
    The garden is shaped like a rectangle.
    The garden is 15 yards wide and 10 yards long.
    How many yards of fence will Mr. James need?

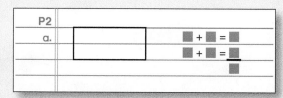

Part 3 Figure out the perimeter of each shape.

a.

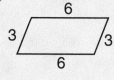

yard

b.

6
3   3
6

centimeter

c.

4
5
3   4
5

feet

P3	
a.	
b.	
c.	

Lesson 127

Part 4 Copy and work each problem.

a. $\begin{array}{r} 450 \\ -\ 99 \end{array}$

b. $\begin{array}{r} 450 \\ +\ 99 \end{array}$

c. $\begin{array}{r} 132 \\ \times\quad 9 \end{array}$

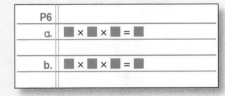

Part 5 Work each problem.

a. The store sold 45 apples and 5 bananas.
   The store sold how many times as many apples as bananas?

b. The oak tree is 15 meters tall.
   The redwood tree is 4 times as tall as the oak tree.
   How many meters tall is the redwood tree?

Part 6 Find the volume of each rectangular prism.

a. 4 in. 5 in. 6 in.

b. 2 in. 3 in. 5 in.

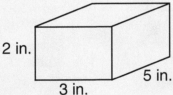

Part 7 Work a fraction problem.

a. A pizza was divided into 10 equal slices.
   7 slices had pineapple.
   What fraction of the pizza did not have pineapple?

Part 8 For each item, make a number family and work the problem.

a. 1 meter equals 100 centimeters.
   How many centimeters is 6 meters?

b. 1 week equals 7 days.
   How many weeks is 140 days?

Lesson 127

Part 9 Write the missing number for each number family.

a. 6 ⟶ 48

b. 7 ⟶ 49

c. 6 →⁹ ___

d. 8 →⁸ ___

e. ___ →⁸ 32

f. 4 →⁸ ___

g. ___ →⁸ 24

h. ___ →⁶ 24

i. ___ →³ 24

j. ___ →⁹ 54

k. ___ →⁸ 72

l. ___ →⁴ 36

m. ___ →⁸ 48

n. 6 →³ ___

o. ___ →⁴ 32

P9		
a.	f.	k.
b.	g.	l.
c.	h.	m.
d.	i.	n.
e.	j.	o.

Lesson 128

Part 1

a. Pam had 97 cups.
She gave 42 cups to her brother and 11 cups to her sister.
How many cups did Pam end up with?

b. Tom had 54 model cars.
He gave 13 to one friend and 18 to another friend.
How many model cars did he end up with?

P1			
a.	▪	b.	▪
	▬		▬
	▪		▪

Independent Work

Part 2 | Work the problem to answer each question.

```
       ┌──────────┬────────┐
       │          │        │
 10 ft │  room A  │ room B │ 10 ft
       │          │        │
       └──────────┴────────┘
          9 ft       6 ft
```

P2	
a.	
b.	
c.	

a. What is the area of room A?

b. What is the area of room B?

c. How many more square feet does room A have than room B has?

Part 3 | Write the missing number in each equation.

a. $5 \times \blacksquare = 20$

b. $2 \times \blacksquare = 12$

P3		
a.		b.

Part 4 | Work each problem by rounding the numbers to the nearest hundred.

a. Jennifer ran 872 meters on Saturday and 614 meters on Sunday.
About how many meters did she run on both days?

b. Tina did 728 math problems in January and 492 math problems in February.
About how many more math problems did Tina do in January than in February?

P4		
a.		b.

Lesson 128

Part 5 Copy and work each problem.

a. 4)200   b. 2)128   c. 6)246

P5			
a.	b.	c.	

Part 6 Work each problem.

a. How much more is 9 dollars than $2.40?

b. How much more is 12 dollars than $8.50?

P6		
a.	b.	

Part 7 Work each problem.

a. Alejandra ran 5 miles every day.
   How many miles did she run in 30 days?

b. All the bags had the same number of apples.
   There were 40 apples in 8 bags.
   How many apples were in each bag?

P7	
a.	
b.	

Part 8 Copy the problems you can work and write the answers.
Don't do anything to the problems you can't work.

P8		
a. $\frac{4}{7} + \frac{2}{7} =$ ___	b. $\frac{5}{9} - \frac{1}{9} =$ ___	c. $\frac{4}{7} + \frac{3}{5} =$ ___

Connecting Math Concepts

Part 9 Write the missing number for each number family.

a. 7 ⌐6→ ___

b. ___ ⌐6→ 54

c. ___ ⌐7→ 28

d. 7 ⌐⌐→ 49

e. 6 ⌐⌐→ 48

f. ___ ⌐4→ 36

g. 8 ⌐⌐→ 64

h. 7 ⌐⌐→ 42

i. 8 ⌐8→ ___

j. ___ ⌐7→ 21

k. ___ ⌐4→ 28

l. 7 ⌐⌐→ 56

m. ___ ⌐8→ 48

n. ___ ⌐9→ 54

o. ___ ⌐6→ 42

P9		
a.	f.	k.
b.	g.	l.
c.	h.	m.
d.	i.	n.
e.	j.	o.

Lesson 129

Part 1

a. Mrs. Jones had 92 paper clips.
She gave 23 paper clips to one class of
students and 28 paper clips to another class.
How many paper clips did she have left?

Bill: 92 + 51 = 143

Mary: 92 − 51 = 41

Jill: 92 − 5 = 87

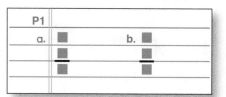

b. The store had 89 cartons of milk.
The store sold 34 cartons in the morning
and 28 cartons in the afternoon.
How many cartons of milk did the store end up with?

Margo: 89 + 34 + 56 = 179

James: 89 − 34 = 55

Fran: 89 − 62 = 27

Independent Work

Part 2 | Work the problem to answer each question.

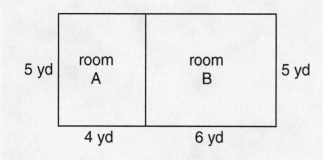

P2	
a.	
b.	
c.	

a. What is the area of room A?

b. What is the area of room B?

c. How many more square yards does
room B have than room A has?

Part 3 | Write the missing number in each equation.

P3		
a.		b.

a. $9 \times \blacksquare = 45$ b. $\blacksquare \times 3 = 15$

Lesson

Part 4 For each item, write the fraction that has the smallest parts.

a. $\dfrac{1}{3}$ $\dfrac{1}{8}$ $\dfrac{1}{5}$ b. $\dfrac{1}{10}$ $\dfrac{1}{5}$ $\dfrac{1}{9}$ c. $\dfrac{1}{6}$ $\dfrac{1}{2}$ $\dfrac{1}{8}$

P4		
a.	b.	c.

Part 5 Copy and work each problem.

a. $\begin{array}{r} 70 \\ \times\ 4 \\ \hline \end{array}$ b. $\begin{array}{r} 242 \\ \times\ \ 9 \\ \hline \end{array}$ c. $\begin{array}{r} 274 \\ \times\ \ 5 \\ \hline \end{array}$

P5		
a. ■	b. ■	c. ■

Part 6 Read the problem. Then answer the questions.

Mrs. Jenkins wants to buy pieces of carpet to cover the floor of a room. The floor is 6 yards wide and 5 yards long.

a. What is the area of the room?

b. How many pieces of carpet does Mrs. Jenkins have to buy if each piece of carpet is 1 square yard?

P6	
a.	b.

Part 7 Copy and work each problem.

a. $\begin{array}{r} 354 \\ -\ 88 \\ \hline \end{array}$ b. $\begin{array}{r} 354 \\ +\ 88 \\ \hline \end{array}$ c. $\begin{array}{r} 354 \\ \times\ \ 8 \\ \hline \end{array}$

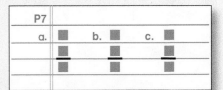

Part 8 Work each problem.

a. There are 125 nails in each package.
The carpenter bought 7 packages of nails.
How many nails did the carpenter buy?

b. Kevin read 4 books every week.
How many weeks will it take Kevin to read 120 books?

c. The horse is 4 times as heavy as the goat.
The horse weighs 840 pounds.
How much does the goat weigh?

P8	
a.	
b.	
c.	

Lesson 129

Part 9 | Write the missing number for each number family.

a. 7 —6→ ___

b. ___ —6→ 54

c. ___ —7→ 21

d. 9 —8→ ___

e. ___ —7→ 28

f. ___ —3→ 24

g. 7 —9→ ___

h. 7 ⟹ 42

i. ___ —4→ 24

j. 7 —7→ ___

k. ___ —4→ 28

l. ___ —9→ 45

m. ___ —9→ 54

n. 9 ⟹ 63

o. ___ —7→ 42

P9		
a.	f.	k.
b.	g.	l.
c.	h.	m.
d.	i.	n.
e.	j.	o.

Lesson

a. Jason had 94 cents.
   He bought a pencil for 18 cents and a notebook for 43 cents.
   How many cents did Jason end up with?

   Delvin: 94 − 43 = 51

   Nestor: 94 − 61 = 153

   Rose: 94 − 61 = 33

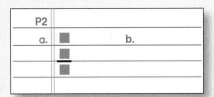

b. The store had 152 bottles of juice.
   The store sold 34 bottles in the morning
   and 56 bottles in the afternoon.
   How many bottles of juice did the store end up with?

   Cristy: 152 + 34 − 56 = 130

   James: 152 − 90 = 62

   Fran: 152 − 90 = 242

Independent Work

Part 2 Read the problem. Then answer the questions.

James wants to buy pieces of tile to cover the floor of the kitchen.
The floor of the kitchen is 21 feet wide and 8 feet long.

a. What is the area of the kitchen?

b. How many pieces of tile does James have to buy
   if each piece of tile is 1 square foot?

Part 3 Write the missing number in each equation.

a. $3 \times \blacksquare = 21$

b. $2 \times \blacksquare = 18$

Lesson 130

Part 4 Copy and work each problem.

a.
$$\begin{array}{r} 4520 \\ -1790 \\ \hline \end{array}$$

b.
$$\begin{array}{r} 4520 \\ +1790 \\ \hline \end{array}$$

c.
$$\begin{array}{r} 452 \\ \times \quad 7 \\ \hline \end{array}$$

P4			
a. ■	b. ■	c. ■	
■	■	■	
■	■	■	

Part 5 Answer both questions.

a. There are two pies that are exactly the same size.
Pie A is divided into 4 slices.
Pie B is divided into 7 slices.
Which pie has the bigger slices?

b. There are two pies that are exactly the same size.
Pie A is divided into 8 slices.
Pie B is divided into 5 slices.
Which pie has the bigger slices?

P5	
a.	b.

Part 6 For each item, make a number family and work the problem.

a. 1 week equals 7 days.
How many weeks is 42 days?

b. 1 meter equals 100 centimeters.
How many centimeters is 8 meters?

c. 1 yard equals 3 feet.
How many feet is 36 yards?

P6	
a.	
b.	
c.	

Lesson 130

Part 7 Write the correct name for the shapes in each box.

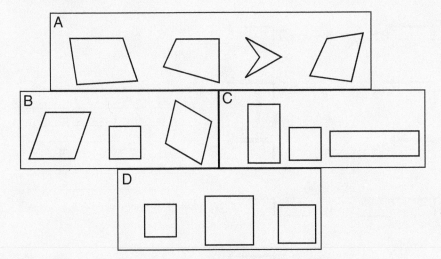

P7	
a.	
b.	
c.	
d.	

rhombus, square, quadrilateral, rectangle

Part 8 Work the problem to answer each question.

This picture shows the rooms in a house.

P8	
a.	
b.	
c.	

10 yd

room A

room B   4 yd

6 yd   5 yd

a. What is the area of room A?

b. What is the area of room B?

c. What is the total area of the house?

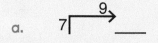

Lesson 130

Part 9 Write the missing number for each number family.

a. 7 ⌐9→ ___

b. ___ ⌐6→ 42

c. ___ ⌐7→ 28

d. ___ ⌐3→ 21

e. ___ ⌐7→ 56

f. ___ ⌐9→ 81

g. ___ ⌐6→ 36

h. 7 ⌐=→ 63

i. 6 ⌐7→ ___

j. ___ ⌐4→ 28

k. 7 ⌐6→ ___

l. 7 ⌐=→ 49

m. 7 ⌐=→ 42

n. ___ ⌐4→ 36

o. ___ ⌐9→ 63

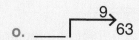

P9		
a.	f.	k.
b.	g.	l.
c.	h.	m.
d.	i.	n.
e.	j.	o.

Connecting Math Concepts

Practice Sets 1-50

Practice Set 1

a. $\begin{array}{r} 4 \\ + 5 \\ \hline \end{array}$ b. $\begin{array}{r} 5 \\ + 7 \\ \hline \end{array}$ c. $\begin{array}{r} 8 \\ + 5 \\ \hline \end{array}$ d. $\begin{array}{r} 5 \\ + 9 \\ \hline \end{array}$ e. $\begin{array}{r} 5 \\ + 8 \\ \hline \end{array}$ f. $\begin{array}{r} 5 \\ + 6 \\ \hline \end{array}$ g. $\begin{array}{r} 7 \\ + 5 \\ \hline \end{array}$ h. $\begin{array}{r} 9 \\ + 5 \\ \hline \end{array}$ i. $\begin{array}{r} 6 \\ + 5 \\ \hline \end{array}$ j. $\begin{array}{r} 5 \\ + 4 \\ \hline \end{array}$

Practice Set 2

a. $\begin{array}{r} 5 \\ + 8 \\ \hline \end{array}$ b. $\begin{array}{r} 5 \\ + 6 \\ \hline \end{array}$ c. $\begin{array}{r} 5 \\ + 9 \\ \hline \end{array}$ d. $\begin{array}{r} 8 \\ + 5 \\ \hline \end{array}$ e. $\begin{array}{r} 7 \\ + 5 \\ \hline \end{array}$ f. $\begin{array}{r} 9 \\ + 5 \\ \hline \end{array}$ g. $\begin{array}{r} 5 \\ + 7 \\ \hline \end{array}$ h. $\begin{array}{r} 6 \\ + 5 \\ \hline \end{array}$ i. $\begin{array}{r} 4 \\ + 5 \\ \hline \end{array}$ j. $\begin{array}{r} 5 \\ + 4 \\ \hline \end{array}$

Practice Set 3

a. $\begin{array}{r} 7 \\ + 5 \\ \hline \end{array}$ b. $\begin{array}{r} 9 \\ + 5 \\ \hline \end{array}$ c. $\begin{array}{r} 8 \\ + 5 \\ \hline \end{array}$ d. $\begin{array}{r} 5 \\ + 7 \\ \hline \end{array}$ e. $\begin{array}{r} 5 \\ + 9 \\ \hline \end{array}$ f. $\begin{array}{r} 6 \\ + 5 \\ \hline \end{array}$ g. $\begin{array}{r} 5 \\ + 6 \\ \hline \end{array}$ h. $\begin{array}{r} 4 \\ + 5 \\ \hline \end{array}$ i. $\begin{array}{r} 5 \\ + 8 \\ \hline \end{array}$ j. $\begin{array}{r} 5 \\ + 4 \\ \hline \end{array}$

Practice Set 4

a. $\begin{array}{r} 5 \\ + 9 \\ \hline \end{array}$ b. $\begin{array}{r} 6 \\ + 5 \\ \hline \end{array}$ c. $\begin{array}{r} 5 \\ + 7 \\ \hline \end{array}$ d. $\begin{array}{r} 5 \\ + 6 \\ \hline \end{array}$ e. $\begin{array}{r} 8 \\ + 5 \\ \hline \end{array}$ f. $\begin{array}{r} 4 \\ + 5 \\ \hline \end{array}$ g. $\begin{array}{r} 9 \\ + 5 \\ \hline \end{array}$ h. $\begin{array}{r} 5 \\ + 8 \\ \hline \end{array}$ i. $\begin{array}{r} 7 \\ + 5 \\ \hline \end{array}$ j. $\begin{array}{r} 5 \\ + 4 \\ \hline \end{array}$

Practice Set 5

a. $\begin{array}{r} 8 \\ + 5 \\ \hline \end{array}$ b. $\begin{array}{r} 4 \\ + 5 \\ \hline \end{array}$ c. $\begin{array}{r} 5 \\ + 6 \\ \hline \end{array}$ d. $\begin{array}{r} 9 \\ + 5 \\ \hline \end{array}$ e. $\begin{array}{r} 5 \\ + 7 \\ \hline \end{array}$ f. $\begin{array}{r} 5 \\ + 8 \\ \hline \end{array}$ g. $\begin{array}{r} 6 \\ + 5 \\ \hline \end{array}$ h. $\begin{array}{r} 7 \\ + 5 \\ \hline \end{array}$ i. $\begin{array}{r} 5 \\ + 9 \\ \hline \end{array}$ j. $\begin{array}{r} 5 \\ + 4 \\ \hline \end{array}$

Practice Set 6

a. $\begin{array}{r} 9 \\ + 8 \\ \hline \end{array}$ b. $\begin{array}{r} 8 \\ + 9 \\ \hline \end{array}$ c. $\begin{array}{r} 8 \\ + 5 \\ \hline \end{array}$ d. $\begin{array}{r} 7 \\ + 8 \\ \hline \end{array}$ e. $\begin{array}{r} 5 \\ + 8 \\ \hline \end{array}$ f. $\begin{array}{r} 8 \\ + 7 \\ \hline \end{array}$ g. $\begin{array}{r} 7 \\ + 7 \\ \hline \end{array}$ h. $\begin{array}{r} 6 \\ + 8 \\ \hline \end{array}$ i. $\begin{array}{r} 4 \\ + 8 \\ \hline \end{array}$ j. $\begin{array}{r} 8 \\ + 6 \\ \hline \end{array}$

Practice Sets 1-50

Practice Set 7

a.	b.	c.	d.	e.	f.	g.	h.	i.	j.
5	7	8	8	7	8	6	9	4	8
+ 8	+ 8	+ 6	+ 5	+ 7	+ 7	+ 8	+ 8	+ 8	+ 9

Practice Set 8

a.	b.	c.	d.	e.	f.	g.	h.	i.	j.
8	7	6	8	9	8	4	7	8	5
+ 6	+ 7	+ 8	+ 5	+ 8	+ 7	+ 8	+ 8	+ 9	+ 8

Practice Set 9

a.	b.	c.	d.	e.	f.	g.	h.	i.	j.
9	8	8	4	6	7	7	8	8	5
+ 8	+ 7	+ 5	+ 8	+ 8	+ 8	+ 7	+ 9	+ 6	+ 8

Practice Set 10

a.	b.	c.	d.	e.	f.	g.	h.	i.	j.
6	5	4	7	7	8	8	8	8	9
+ 8	+ 8	+ 8	+ 8	+ 7	+ 5	+ 9	+ 7	+ 6	+ 8

Practice Set 11

a.	b.	c.	d.	e.	f.	g.	h.	i.	j.
12	14	14	13	11	13	12	16	14	14
− 5	− 9	− 5	− 8	− 5	− 5	− 7	− 8	− 8	− 6

Practice Set 12

a.	b.	c.	d.	e.	f.	g.	h.	i.	j.
13	14	12	11	13	12	11	14	16	14
− 8	− 5	− 7	− 5	− 5	− 5	− 6	− 8	− 8	− 6

Practice Sets 1-50

Practice Set 13

a. 11 − 6 b. 13 − 5 c. 14 − 9 d. 11 − 5 e. 14 − 5 f. 13 − 8 g. 12 − 5 h. 14 − 6 i. 14 − 8 j. 16 − 8

Practice Set 14

a. 12 − 7 b. 14 − 5 c. 11 − 5 d. 13 − 8 e. 12 − 5 f. 13 − 5 g. 11 − 6 h. 16 − 8 i. 14 − 6 j. 14 − 8

Practice Set 15

a. 11 − 5 b. 12 − 5 c. 12 − 7 d. 13 − 5 e. 14 − 5 f. 13 − 8 g. 14 − 9 h. 14 − 6 i. 14 − 8 j. 16 − 8

Practice Set 16

a. 14 − 8 b. 14 − 6 c. 16 − 8 d. 13 − 8 e. 17 − 9 f. 12 − 7 g. 15 − 8 h. 14 − 5 i. 11 − 5 j. 13 − 8

Practice Set 17

a. 17 − 9 b. 14 − 8 c. 13 − 8 d. 12 − 5 e. 12 − 7 f. 13 − 5 g. 17 − 8 h. 14 − 6 i. 15 − 7 j. 16 − 8

Practice Set 18

a. 15 − 7 b. 13 − 8 c. 14 − 6 d. 17 − 8 e. 12 − 5 f. 15 − 8 g. 17 − 9 h. 13 − 5 i. 14 − 8 j. 16 − 8

Practice Sets 1-50

Practice Set 19

a.	b.	c.	d.	e.	f.	g.	h.	i.	j.
14	13	17	13	15	14	16	15	13	11
− 9	− 5	− 8	− 8	− 8	− 8	− 8	− 7	− 5	− 6

Practice Set 20

a.	b.	c.	d.	e.	f.	g.	h.	i.	j.
13	14	15	12	14	17	13	12	16	17
− 8	− 5	− 8	− 5	− 8	− 8	− 5	− 7	− 8	− 9

Practice Set 21

a.	b.	c.	d.	e.	f.	g.	h.	i.	j.
8	5	8	5	9	8	5	6	7	8
+ 9	+ 6	+ 7	+ 7	+ 8	+ 6	+ 9	+ 8	+ 9	+ 8

Practice Set 22

a.	b.	c.	d.	e.	f.	g.	h.	i.	j.
9	8	5	5	8	6	5	7	8	8
+ 8	+ 6	+ 7	+ 9	+ 8	+ 8	+ 6	+ 9	+ 9	+ 7

Practice Set 23

a.	b.	c.	d.	e.	f.	g.	h.	i.	j.
8	6	5	5	5	7	8	8	9	8
+ 8	+ 8	+ 9	+ 6	+ 7	+ 9	+ 6	+ 7	+ 8	+ 9

Practice Set 24

a.	b.	c.	d.	e.	f.	g.	h.	i.	j.
7	8	5	8	5	5	8	7	9	8
+ 8	+ 9	+ 6	+ 7	+ 9	+ 7	+ 8	+ 9	+ 5	+ 6

Connecting Math Concepts

Practice Sets 1-50

Practice Set 25

a. 8
+ 8

b. 7
+ 9

c. 5
+ 9

d. 9
+ 7

e. 5
+ 7

f. 9
+ 5

g. 8
+ 6

h. 5
+ 6

i. 8
+ 9

j. 8
+ 7

Practice Set 26

a. 5
+ 6

b. 8
+ 8

c. 6
+ 6

d. 6
+ 5

e. 7
+ 6

f. 8
+ 7

g. 9
+ 6

h. 7
+ 8

i. 6
+ 9

j. 6
+ 7

Practice Set 27

a. 7
+ 6

b. 6
+ 5

c. 8
+ 7

d. 9
+ 6

e. 6
+ 6

f. 7
+ 8

g. 8
+ 8

h. 5
+ 6

i. 6
+ 7

j. 8
+ 9

Practice Set 28

a. 6
+ 6

b. 9
+ 6

c. 8
+ 7

d. 6
+ 5

e. 7
+ 6

f. 6
+ 9

g. 6
+ 7

h. 5
+ 6

i. 8
+ 8

j. 7
+ 8

Practice Set 29

a. 6
+ 9

b. 7
+ 6

c. 6
+ 7

d. 6
+ 5

e. 5
+ 6

f. 8
+ 6

g. 8
+ 8

h. 9
+ 6

i. 6
+ 8

j. 6
+ 6

Practice Set 30

a. 5
+ 6

b. 8
+ 6

c. 6
+ 5

d. 8
+ 8

e. 6
+ 7

f. 9
+ 6

g. 7
+ 6

h. 6
+ 6

i. 6
+ 8

j. 6
+ 9

Practice Sets

a.   9 b.   9 c.   6 d.   7 e.   9 f.   7 g.   5 h.   6 i.   9 j.   5
  + 6   + 5   + 7   + 9   + 7   + 6   + 8   + 9   + 9   + 9

Practice Set 32

a.   8 b.   5 c.   7 d.   9 e.   6 f.   9 g.   7 h.   6 i.   9 j.   9
  + 5   + 9   + 6   + 5   + 9   + 9   + 5   + 7   + 6   + 7

Practice Set 33

a.   9 b.   7 c.   9 d.   5 e.   9 f.   8 g.   6 h.   9 i.   7 j.   6
  + 5   + 9   + 6   + 9   + 7   + 6   + 9   + 9   + 6   + 7

Practice Set 34

a.   6 b.   9 c.   6 d.   9 e.   5 f.   7 g.   9 h.   9 i.   6 j.   5
  + 5   + 9   + 7   + 6   + 8   + 6   + 7   + 5   + 9   + 9

Practice Set 35

a.   6 b.   9 c.   9 d.   9 e.   6 f.   8 g.   6 h.   7 i.   5 j.   9
  + 7   + 5   + 6   + 9   + 8   + 9   + 9   + 6   + 9   + 7

Practice Set 36

a. 15 b. 11 c. 14 d. 12 e. 15 f. 13 g. 15 h. 13 i. 11 j. 14
  − 6   − 5   − 6   − 6   − 8   − 6   − 7   − 7   − 6   − 8

Practice Sets 1-50

Practice Set 37

a. 15 b. 11 c. 14 d. 15 e. 14 f. 12 g. 15 h. 15 i. 13 j. 13
   − 9   − 6   − 6   − 8   − 8   − 6   − 6   − 7   − 6   − 7

Practice Set 38

a. 11 b. 15 c. 15 d. 13 e. 11 f. 14 g. 14 h. 12 i. 13 j. 15
   − 5   − 8   − 6   − 7   − 6   − 6   − 8   − 6   − 6   − 7

Practice Set 39

a. 14 b. 15 c. 14 d. 15 e. 11 f. 13 g. 15 h. 12 i. 11 j. 15
   − 8   − 6   − 6   − 7   − 5   − 7   − 8   − 6   − 6   − 9

Practice Set 40

a. 11 b. 13 c. 14 d. 15 e. 14 f. 15 g. 12 h. 11 i. 15 j. 13
   − 6   − 7   − 6   − 6   − 8   − 8   − 6   − 5   − 7   − 6

Practice Set 41

a. 18 b. 13 c. 16 d. 14 e. 17 f. 15 g. 15 h. 16 i. 14 j. 17
   − 9   − 6   − 9   − 5   − 9   − 6   − 9   − 7   − 9   − 8

Practice Set 42

a. 17 b. 14 c. 15 d. 16 e. 15 f. 13 g. 17 h. 14 i. 18 j. 15
   − 9   − 5   − 6   − 9   − 9   − 6   − 8   − 9   − 9   − 9

Practice Sets 1-50

Practice Set 43

a. $\begin{array}{r} 15 \\ -\ 9 \\ \hline \end{array}$ b. $\begin{array}{r} 17 \\ -\ 8 \\ \hline \end{array}$ c. $\begin{array}{r} 16 \\ -\ 9 \\ \hline \end{array}$ d. $\begin{array}{r} 18 \\ -\ 9 \\ \hline \end{array}$ e. $\begin{array}{r} 16 \\ -\ 7 \\ \hline \end{array}$ f. $\begin{array}{r} 17 \\ -\ 9 \\ \hline \end{array}$ g. $\begin{array}{r} 15 \\ -\ 6 \\ \hline \end{array}$ h. $\begin{array}{r} 14 \\ -\ 9 \\ \hline \end{array}$ i. $\begin{array}{r} 15 \\ -\ 6 \\ \hline \end{array}$ j. $\begin{array}{r} 14 \\ -\ 5 \\ \hline \end{array}$

Practice Set 44

a. $\begin{array}{r} 13 \\ -\ 6 \\ \hline \end{array}$ b. $\begin{array}{r} 18 \\ -\ 9 \\ \hline \end{array}$ c. $\begin{array}{r} 14 \\ -\ 6 \\ \hline \end{array}$ d. $\begin{array}{r} 17 \\ -\ 9 \\ \hline \end{array}$ e. $\begin{array}{r} 15 \\ -\ 6 \\ \hline \end{array}$ f. $\begin{array}{r} 16 \\ -\ 9 \\ \hline \end{array}$ g. $\begin{array}{r} 16 \\ -\ 7 \\ \hline \end{array}$ h. $\begin{array}{r} 15 \\ -\ 9 \\ \hline \end{array}$ i. $\begin{array}{r} 17 \\ -\ 8 \\ \hline \end{array}$ j. $\begin{array}{r} 14 \\ -\ 9 \\ \hline \end{array}$

Practice Set 45

a. $\begin{array}{r} 15 \\ -\ 6 \\ \hline \end{array}$ b. $\begin{array}{r} 16 \\ -\ 9 \\ \hline \end{array}$ c. $\begin{array}{r} 17 \\ -\ 9 \\ \hline \end{array}$ d. $\begin{array}{r} 16 \\ -\ 7 \\ \hline \end{array}$ e. $\begin{array}{r} 14 \\ -\ 6 \\ \hline \end{array}$ f. $\begin{array}{r} 15 \\ -\ 9 \\ \hline \end{array}$ g. $\begin{array}{r} 18 \\ -\ 9 \\ \hline \end{array}$ h. $\begin{array}{r} 14 \\ -\ 9 \\ \hline \end{array}$ i. $\begin{array}{r} 13 \\ -\ 6 \\ \hline \end{array}$ j. $\begin{array}{r} 17 \\ -\ 8 \\ \hline \end{array}$

Practice Set 46

a. $\begin{array}{r} 15 \\ -\ 8 \\ \hline \end{array}$ b. $\begin{array}{r} 16 \\ -\ 9 \\ \hline \end{array}$ c. $\begin{array}{r} 13 \\ -\ 7 \\ \hline \end{array}$ d. $\begin{array}{r} 13 \\ -\ 4 \\ \hline \end{array}$ e. $\begin{array}{r} 13 \\ -\ 6 \\ \hline \end{array}$ f. $\begin{array}{r} 12 \\ -\ 5 \\ \hline \end{array}$ g. $\begin{array}{r} 15 \\ -\ 7 \\ \hline \end{array}$ h. $\begin{array}{r} 12 \\ -\ 7 \\ \hline \end{array}$ i. $\begin{array}{r} 17 \\ -\ 8 \\ \hline \end{array}$ j. $\begin{array}{r} 13 \\ -\ 5 \\ \hline \end{array}$

Practice Set 47

a. $\begin{array}{r} 13 \\ -\ 4 \\ \hline \end{array}$ b. $\begin{array}{r} 15 \\ -\ 7 \\ \hline \end{array}$ c. $\begin{array}{r} 12 \\ -\ 5 \\ \hline \end{array}$ d. $\begin{array}{r} 15 \\ -\ 8 \\ \hline \end{array}$ e. $\begin{array}{r} 12 \\ -\ 7 \\ \hline \end{array}$ f. $\begin{array}{r} 13 \\ -\ 6 \\ \hline \end{array}$ g. $\begin{array}{r} 16 \\ -\ 9 \\ \hline \end{array}$ h. $\begin{array}{r} 17 \\ -\ 8 \\ \hline \end{array}$ i. $\begin{array}{r} 13 \\ -\ 5 \\ \hline \end{array}$ j. $\begin{array}{r} 13 \\ -\ 7 \\ \hline \end{array}$

Practice Set 48

a. $\begin{array}{r} 13 \\ -\ 7 \\ \hline \end{array}$ b. $\begin{array}{r} 17 \\ -\ 8 \\ \hline \end{array}$ c. $\begin{array}{r} 16 \\ -\ 9 \\ \hline \end{array}$ d. $\begin{array}{r} 12 \\ -\ 7 \\ \hline \end{array}$ e. $\begin{array}{r} 13 \\ -\ 4 \\ \hline \end{array}$ f. $\begin{array}{r} 13 \\ -\ 6 \\ \hline \end{array}$ g. $\begin{array}{r} 15 \\ -\ 8 \\ \hline \end{array}$ h. $\begin{array}{r} 15 \\ -\ 7 \\ \hline \end{array}$ i. $\begin{array}{r} 12 \\ -\ 5 \\ \hline \end{array}$ j. $\begin{array}{r} 17 \\ -\ 8 \\ \hline \end{array}$

Practice Sets 1-50

a. 13 b. 13 c. 15 d. 15 e. 13 f. 12 g. 17 h. 16 i. 13 j. 12
$-\ 6$ $-\ 4$ $-\ 8$ $-\ 7$ $-\ 7$ $-\ 5$ $-\ 8$ $-\ 9$ $-\ 5$ $-\ 7$

a. 12 b. 15 c. 16 d. 13 e. 17 f. 13 g. 15 h. 13 i. 13 j. 12
$-\ 5$ $-\ 7$ $-\ 9$ $-\ 6$ $-\ 8$ $-\ 4$ $-\ 8$ $-\ 5$ $-\ 7$ $-\ 7$

Level D Correlation to Grade 3
Common Core State Standards for Mathematics

Operations and Algebraic Thinking (3.OA)

Represent and solve problems involving multiplication and division.

1. Interpret products of whole numbers, e.g., interpret 5 × 7 as the total number of objects in 5 groups of 7 objects each ***or 7 groups of 5 objects each.** *For example, describe a context in which a total number of objects can be expressed as 5 × 7.*

Lessons	WB: 5–8, 10–12, 14–16, 24, 31, 128 TB: 16–20, 22, 26, 29, 31, 33, 58–60, 72–77, 79–84, 90–93, 95–97, 99, 104–115, 117–122, 124, 125, 129

*Denotes California-only content.

Operations and Algebraic Thinking (3.OA)

Represent and solve problems involving multiplication and division.

2. Interpret whole-number quotients of whole numbers, e.g., interpret 56 ÷ 8 as the number of objects in each share when 56 objects are partitioned equally into 8 shares, or as a number of shares when 56 objects are partitioned into equal shares of 8 objects each. *For example, describe a context in which a number of shares or a number of groups can be expressed as 56 ÷ 8.*

Lessons	TB: 73, 75, 77–79, 85, 87, 88, 90, 92, 94, 95, 97, 99, 104, 107–113, 115–119, 121, 128–130

Operations and Algebraic Thinking (3.OA)

Represent and solve problems involving multiplication and division.

3. Use multiplication and division within 100 to solve word problems in situations involving equal groups, arrays, and measurement quantities, e.g., by using drawings and equations with a symbol for the unknown number to represent the problem.

Lessons	TB: 72–125, 127, 129

Operations and Algebraic Thinking (3.OA)

Represent and solve problems involving multiplication and division.

4. Determine the unknown whole number in a multiplication or division equation relating three whole numbers. *For example, determine the unknown number that makes the equation true in each of the equations 8 × ? = 48, 5 = ■ ÷ 3, 6 × 6 = ?*

Lessons	WB: 5–8, 10, 24, 31, 32, 34–43, 46–55, 57–70, 71–98, 100–130 TB: 54, 55, 59, 66–102, 104–130 Student Practice Software: Block 3 Activity 2

Operations and Algebraic Thinking (3.OA)

Understand properties of multiplication and the relationship between multiplication and division.

5. Apply properties of operations as strategies to multiply and divide. *Examples: If 6 × 4 = 24 is known, then 4 × 6 = 24 is also known. (Commutative property of multiplication.) 3 × 5 × 2 can be found by 3 × 5 = 15, then 15 × 2 = 30, or by 5 × 2 = 10, then 3 × 10 = 30. (Associative property of multiplication.) Knowing that 8 × 5 = 40 and 8 × 2 = 16, one can find 8 × 7 as 8 × (5 + 2) = (8 × 5) + (8 × 2) = 40 + 16 = 56. (Distributive property.)*

Lessons	WB: 29–34, 37, 48, 57, 64–67, 69–72, 75–77, 84–87, 91–94, 101–107, 113, 114, 117–128, 130 TB: 55, 56 Student Practice Software: Block 5 Activities 1 and 4

Operations and Algebraic Thinking (3.OA)

Multiply and divide within 100.

7. Fluently multiply and divide within 100, using strategies such as the relationship between multiplication and division (e.g., knowing that 8 × 5 = 40, one knows 40 ÷ 5 = 8) or properties of operations. By the end of Grade 3, know from memory all products of two one-digit numbers.

Lessons	WB: 31–130 TB: 39, 54, 55, 57–61, 66–69, 71–82, 85–88, 90, 91, 93, 98–104, 106–120, 122–130 Student Practice Software: Block 2 Activity 2, Block 3 Activity 3, Block 4 Activity 4

Operations and Algebraic Thinking (3.OA)

Solve problems involving the four operations, and identify and explain patterns in arithmetic.

8. Solve two-step word problems using the four operations. Represent these problems using equations with a letter standing for the unknown quantity. Assess the reasonableness of answers using mental computation and estimation strategies including rounding.

Lessons	TB: 111, 112, 116–125, 128–130

Operations and Algebraic Thinking (3.OA)

Solve problems involving the four operations, and identify and explain patterns in arithmetic.

9. Identify arithmetic patterns (including patterns in the addition table or multiplication table), and explain them using properties of operations. *For example, observe that 4 times a number is always even, and explain why 4 times a number can be decomposed into two equal addends.*

Lessons	WB: 3, 6, 8–14, 21, 22, 26, 31, 32, 38, 42, 45, 65, 66, 87–89, 94–96, 99–101, 113, 114, 118, 129 TB: 17–19, 26, 41, 42, 51–53, 55, 61, 65, 70, 72, 73, 75, 81, 83, 86, 91–93, 97, 98, 100, 101, 113, 117, 119, 120 Student Practice Software: Block 1 Activity 4

Number and Operations in Base Ten (3.NBT)

Use place value understanding and properties of operations to perform multi-digit arithmetic.

1. Use place value understanding to round whole numbers to the nearest 10 or 100.

Lessons	WB: 26–28, 30–34, 36, 37, 56–58, 105, 106, 110–112 TB: 38, 40, 42, 48, 49, 59–72, 74, 75, 77, 79, 80, 81, 94, 96, 98–108, 112–118, 120–123, 128–130 Student Practice Software: Block 2 Activity 5

Number and Operations in Base Ten (3.NBT)

Use place value understanding and properties of operations to perform multi-digit arithmetic.

2. Fluently add and subtract within 1000 using strategies and algorithms based on place value, properties of operations, and/or the relationship between addition and subtraction.

Lessons	WB: 1–51, 55, 58, 61, 62, 71, 77, 86, 108, 127, 129, 130 TB: 15–101, 103–106, 108–130 Student Practice Software: Block 1 Activities 1–3, Block 2 Activities 4 and 5, Block 4 Activity 4

Number and Operations in Base Ten (3.NBT)

Use place value understanding and properties of operations to perform multi-digit arithmetic.

3. Multiply one-digit whole numbers by multiples of 10 in the range 10–90 (e.g., 9×80, 5×60) using strategies based on place value and properties of operations.

Lessons	WB: 5–8, 13, 15, 24, 29, 31–35, 41–45, 53, 78, 129 TB: 16, 71, 72, 74, 76, 78–80, 82, 94, 91–96, 98–108, 112–114, 118, 119, 121, 122, 126, 128, 130

Number and Operations—Fractions (3.NF)

Develop understanding of fractions as numbers.

1. Understand a fraction 1/b as the quantity formed by 1 part when a whole is partitioned into b equal parts; understand a fraction a/b as the quantity formed by a parts of size 1/b.

Lessons	WB: 31–35, 40, 47–49, 53, 55–57, 59, 101–107, 109, 110, 114, 118–120, 126–130 TB: 36–38, 40, 42–46, 48, 51, 52, 54, 56, 61–63, 65, 69, 70, 72, 76, 77, 81, 82, 86, 89, 91, 93, 95, 98, 103, 104, 107, 111, 112, 116, 117, 122–125, 129, 130 **Student Practice Software: Block 2 Activities 3 and 6, Block 3 Activity 4, Block 5 Activities 2 and 5**

Number and Operations—Fractions (3.NF)

Develop understanding of fractions as numbers.

2. Understand a fraction as a number on the number line; represent fractions on a number line diagram.
   a. Represent a fraction 1/b on a number line diagram by defining the interval from 0 to 1 as the whole and partitioning it into b equal parts. Recognize that each part has size 1/b and that the endpoint of the part based at 0 locates the number 1/b on the number line.
   b. Represent a fraction a/b on a number line diagram by marking off a lengths 1/b from 0. Recognize that the resulting interval has size a/b and that its endpoint locates the number a/b on the number line.

Lessons	WB: 43, 47–49, 52–57, 59, 101–107, 109, 110, 114, 126, 129, 130 TB: 44–46, 51, 61–63, 69, 70, 72, 76, 77, 80, 82, 86, 93, 95, 98, 112, 113, 115–123 **Student Practice Software: Block 2 Activity 3, Block 3 Activity 4**

Number and Operations—Fractions (3.NF)

Develop understanding of fractions as numbers.

3. Explain equivalence of fractions in special cases, and compare fractions by reasoning about their size.
   a. Understand two fractions as equivalent (equal) if they are the same size, or the same point on a number line.
   b. Recognize and generate simple equivalent fractions, e.g., 1/2 = 2/4, 4/6 = 2/3). Explain why the fractions are equivalent, e.g., by using a visual fraction model.
   c. Express whole numbers as fractions, and recognize fractions that are equivalent to whole numbers. *Examples: Express 3 in the form 3 = 3/1; recognize that 6/1 = 6; locate 4/4 and 1 at the same point of a number line diagram.*
   d. Compare two fractions with the same numerator or the same denominator by reasoning about their size. Recognize that comparisons are valid only when the two fractions refer to the same whole. Record the results of comparisons with the symbols >, =, or <, and justify the conclusions, e.g., by using a visual fraction model.

Lessons	WB: 34–43, 46, 48, 52–57, 59, 60, 62–65, 72–76, 90–98, 101–107, 109, 110, 114, 119, 120 TB: 43–45, 53, 57, 60–63, 66, 67, 69–71, 77–79, 82, 83, 85–87, 89, 90, 94, 98–100, 102, 104, 106, 108, 110, 112, 114 **Student Practice Software: Block 3 Activity 1, Block 4 Activity 2, Block 5 Activities 3 and 6**

Measurement and Data (3.MD)

Solve problems involving measurement and estimation of intervals of time, liquid volumes, and masses of objects.

1. Tell and write time to the nearest minute and measure time intervals in minutes. Solve word problems involving addition and subtraction of time intervals in minutes, e.g., by representing the problem on a number line diagram.

Lessons	WB: 54
	TB: 55–60, 62–68, 70–75, 77, 80, 82, 84, 85, 88, 90, 92–97, 99–106, 111, 116, 119

Measurement and Data (3.MD)

Solve problems involving measurement and estimation of intervals of time, liquid volumes, and masses of objects.

2. Measure and estimate liquid volumes and masses of objects using standard units of grams (g), kilograms (kg), and liters (l). Add, subtract, multiply, or divide to solve one-step word problems involving masses or volumes that are given in the same units, e.g., by using drawings (such as a beaker with a measurement scale) to represent the problem.

Lessons	WB: 16, 121–123, 129, 130
	TB: 19, 20, 22–25, 27, 32, 33, 40, 42–44, 46, 48, 50, 51, 58, 60, 65, 66, 72, 73, 78, 88–94, 97, 99–101, 104, 107, 110, 112, 115, 116, 118–125, 127, 128

Measurement and Data (3.MD)

Represent and interpret data.

3. Draw a scaled picture graph and a scaled bar graph to represent a data set with several categories. Solve one- and two-step "how many more" and "how many less" problems using information presented in scaled bar graphs. *For example, draw a bar graph in which each square in the bar graph might represent 5 pets.*

Lessons	WB: 75, 76, 124–126, 129, 130
	TB: 63–66, 71–75, 77–99, 105, 106, 108, 110, 111, 113, 115, 120, 124
	Student Practice Software: Block 3 Activity 5, Block 4 Activity 5

Measurement and Data (3.MD)

Represent and interpret data.

4. Generate measurement data by measuring lengths using rulers marked with halves and fourths of an inch. Show the data by making a line plot, where the horizontal scale is marked off in appropriate units—whole numbers, halves, or quarters.

Lessons	WB: 125–130

Measurement and Data (3.MD)

Geometric measurement: understand concepts of area and relate area to multiplication and to addition.

5. Recognize area as an attribute of plane figures and understand concepts of area measurement.
   a. A square with side length 1 unit, called "a unit square," is said to have "one square unit" of area, and can be used to measure area.
   b. A plane figure which can be covered without gaps or overlaps by *n* unit squares is said to have an area of *n* square units.

Lessons	WB: 129 TB: 26, 37, 38, 125 **Student Practice Software: Block 3 Activity 6, Block 4 Activity 6**

Measurement and Data (3.MD)

Geometric measurement: understand concepts of area and relate area to multiplication and to addition.

6. Measure areas by counting unit squares (square cm, square m, square in, square ft, and improvised units).

Lessons	TB: 26, 92, 95 **Student Practice Software: Block 3 Activity 6, Block 4 Activity 6**

Measurement and Data (3.MD)

Geometric measurement: understand concepts of area and relate area to multiplication and to addition.

7. Relate area to the operations of multiplication and addition.
   a. Find the area of a rectangle with whole-number side lengths by tiling it, and show that the area is the same as would be found by multiplying the side lengths.
   b. Multiply side lengths to find areas of rectangles with whole-number side lengths in the context of solving real-world and mathematical problems, and represent whole-number products as rectangular areas in mathematical reasoning.
   c. Use tiling to show in a concrete case that the area of a rectangle with whole-number side lengths a and $b + c$ is the sum of $a \times b$ and $a \times c$. Use area models to represent the distributive property in mathematical reasoning.
   d. Recognize area as additive. Find areas of rectilinear figures by decomposing them into non-overlapping rectangles and adding the areas of the non-overlapping parts, applying this technique to solve real world problems.

Lessons	WB: 35, 129, 130 TB: 37, 39–43, 45, 47, 50–52, 63, 65, 68, 71, 82, 83, 89, 94, 97, 100, 102–104, 106–108, 111, 112, 115, 117, 120, 123–126, 128–130 **Student Practice Software: Block 3 Activity 6, Block 4 Activity 6, Block 5 Activities 1 and 4**

Measurement and Data (3.MD)

Geometric measurement: recognize perimeter as an attribute of plane figures and distinguish between linear and area measures.

8. Solve real world and mathematical problems involving perimeters of polygons, including finding the perimeter given the side lengths, finding an unknown side length, and exhibiting rectangles with the same perimeter and different areas or with the same area and different perimeters.

Lessons	WB: 27–30, 32, 34, 35
	TB: 26, 37, 38, 40–43, 45, 47, 50–52, 63, 65, 68, 71, 82, 83, 89, 91, 94, 97, 100, 106, 107, 115, 117–127

Geometry (3.G)

Reason with shapes and their attributes.

1. Understand that shapes in different categories (e.g., rhombuses, rectangles, and others) may share attributes (e.g., having four sides), and that the shared attributes can define a larger category (e.g., quadrilaterals). Recognize rhombuses, rectangles, and squares as examples of quadrilaterals, and draw examples of quadrilaterals that do not belong to any of these subcategories.

Lessons	WB: 27–29, 81, 129, 130
	TB: 31, 33–36, 39, 41, 45, 46, 59, 63, 76, 82–85, 87, 101, 130
	Student Practice Software: Block 4 Activity 3

Geometry (3.G)

Reason with shapes and their attributes.

2. Partition shapes into parts with equal areas. Express the area of each part as a unit fraction of the whole. *For example, partition a shape into 4 parts with equal area, and describe the area of each part as 1/4 of the area of the shape.*

Lessons	WB: 121
	Student Practice Software: Block 5 Activities 2 and 5

Standards for Mathematical Practice

Connecting Math Concepts addresses all of the Standards for Mathematical Practice throughout the program. What follows are examples of how individual standards are addressed in this level.

1. Make sense of problems and persevere in solving them.

Word Problems (Lessons 9–52, 59–108): Students learn to identify specific types of word problems (i.e., comparison, start-end, discrimination) and set up and solve the problems based on the specific problem types.

2. Reason abstractly and quantitatively.

Foundation—Counting (Lessons 1–21, 59–63): Beginning in Lesson 13, students work "count-by" multiplication problems to find the number of squares in rectangular arrays, understanding the relationship between the digits in their written work and the number of rows in an array and the number of squares in each row.

3. Construct viable arguments and critique the reasoning of others.

Estimation (Lessons 22–26, 34–60, 96–115, 116–130): Students learn how to round numbers and then apply that knowledge to word problems involving estimation. They work the original problem and the estimation problem and then compare answers to verify that the estimated answer is close to the exact answer. Students can construct an argument to persuade someone whether an estimated answer is reasonable.

4. Model with mathematics.

Number Families (Lessons 1–20, 29–45, 64–72, 77–130): Students learn to represent three related numbers in a number family. Later, they apply the number-family strategy to model and solve word problems.

5. Use appropriate tools strategically.

Throughout the program (Lessons 1–130) students use pencils, workbooks, lined paper, and textbooks to complete their work. They use rulers to draw lines that divide shapes into equal parts. They use the computer to access the Practice Software where they apply the skills they learn in the lessons.

6. Attend to precision.

Geometry (Lessons 12–34, 76–87, 106–118): When finding area, perimeter, and volume, students learn to attend to the units and are required to respond verbally and in written answers with the correct unit.

7. Look for and make use of structure.

Facts (Lessons 31–130): In Lesson 106, students first learn about the associative law of multiplication. In Lesson 108, they first learn about the commutative property of addition. They use the laws to add and multiply three numbers.

8. Look for and express regularity in repeated reasoning.

Place Value and Column Problems (Lessons 1–25): Students first learn to rename 2-digit numbers using knowledge of place value. For example, they rewrite 62 as 50 + 12. This skill is applied to the renaming procedure used in column problems. In computing 52 − 19, students rewrite 52 as 4 tens plus 12. The skill is applied again to column problems involving 3-digit numbers. In computing 502 − 19, they rewrite 502 as 49 tens plus 12.

Photo Credits